Learning and Decision-Making from Rank Data

Synthesis Lectures on Artificial Intelligence and Machine Learning

Editors

Ronald J. Brachman, *Jacobs Technion-Cornell Institute at Cornell Tech*
Francesca Rossi, *IBM Research AI*
Peter Stone, *University of Texas at Austin*

Algorithms for Reinforcement Learning
Csaba Szepesvári
2010

Data Integration: The Relational Logic Approach
Michael Genesereth
2010

Markov Logic: An Interface Layer for Artificial Intelligence
Pedro Domingos and Daniel Lowd
2009

Introduction to Semi-Supervised Learning
XiaojinZhu and Andrew B.Goldberg
2009

Action Programming Languages
Michael Thielscher
2008

Representation Discovery using Harmonic Analysis
Sridhar Mahadevan
2008

Essentials of Game Theory: A Concise Multidisciplinary Introduction
Kevin Leyton-Brown and Yoav Shoham
2008

A Concise Introduction to Multiagent Systems and Distributed Artificial Intelligence
Nikos Vlassis
2007

Intelligent Autonomous Robotics: A Robot Soccer Case Study
Peter Stone
2007

Learning and Decision-Making from Rank Data

Lirong Xia

ISBN: 978-3-031-00454-4   paperback
ISBN: 978-3-031-01582-3   ebook
ISBN: 978-3-031-00027-0   hardcover

DOI 10.1007/978-3-031-01582-3

A Publication in the Springer series
SYNTHESIS LECTURES ON ARTIFICIAL INTELLIGENCE AND MACHINE LEARNING

Lecture #40
Series Editors: Ronald J. Brachman, *Jacobs Technion_Cornell Institute at Cornell Tech*
                Francesca Rossi, *IBM Research AI*
                Peter Stone, *University of Texas at Austin*
Series ISSN
Print 1939-4608   Electronic 1939-4616

Learning and Decision-Making from Rank Data

Lirong Xia
Rensselaer Polytechnic Institute

SYNTHESIS LECTURES ON ARTIFICIAL INTELLIGENCE AND MACHINE LEARNING #40

ABSTRACT

The ubiquitous challenge of learning and decision-making from rank data arises in situations where intelligent systems collect preference and behavior data from humans, learn from the data, and then use the data to help humans make efficient, effective, and timely decisions. Often, such data are represented by *rankings*.

This book surveys some recent progress toward addressing the challenge from the considerations of statistics, computation, and socio-economics. We will cover classical statistical models for rank data, including random utility models, distance-based models, and mixture models. We will discuss and compare classical and state-of-the-art algorithms, such as algorithms based on Minorize-Majorization (MM), Expectation-Maximization (EM), Generalized Method-of-Moments (GMM), rank breaking, and tensor decomposition. We will also introduce principled Bayesian preference elicitation frameworks for collecting rank data. Finally, we will examine socio-economic aspects of statistically desirable decision-making mechanisms, such as Bayesian estimators.

This book can be useful in three ways: (1) for theoreticians in statistics and machine learning to better understand the considerations and caveats of learning from rank data, compared to learning from other types of data, especially cardinal data; (2) for practitioners to apply algorithms covered by the book for sampling, learning, and aggregation; and (3) as a textbook for graduate students or advanced undergraduate students to learn about the field.

This book requires that the reader has basic knowledge in probability, statistics, and algorithms. Knowledge in social choice would also help but is not required.

KEYWORDS

rank data, decision-making, random utility models, Plackett-Luce model, distance-based models, Mallows' model, preference elicitation, social choice, fairness, Bayesian estimators

Contents

Preface

We are living in a world of data. Every day, the booming IT industry collects large amounts of user preference and behavioral data to make various decisions. In many cases, these data are in various rank formats instead of numerical values over alternatives—for example, how voters rank candidates, consumers choose one product over another, search engines rank webpages, etc.

Efficiently collecting and learning from rank data are important challenges to artificial intelligence and machine learning. Recently, there has been a growing literature on preference learning and decision-making that leverages (1) classical statistical modeling for rank data, such as discrete choice models and random utility models, and (2) modern machine learning and optimization techniques. In addition, as the data are often collected from humans and the decisions are made for humans, socio-economic aspects such as fairness are also important considerations.

Most of the book aims to answer the question of how we can design fast and accurate algorithms for learning from rank data in terms of widely studied and applied statistical models. After introducing random utility models and distance-based models in Chapter 2, we discuss several classical and state-of-the-art parameter estimation algorithms in Chapter 3. Chapter 4 covers a general framework for handling rank data called rank-breaking. Chapter 5 focuses on the theory and algorithms of mixture models for rank data.

How to efficiently elicit preferences from users is the main topic of Chapter 6, where we take a Bayesian experimental design and active learning approach by computing the most cost-effective queries. This chapter also covers some Bayesian approaches toward learning from partial orders. Preliminary experiments show that choice data are not as cost-effective as other types of rank data.

Chapter 7 covers a statistical decision-theoretic framework for social choice, under which tradeoffs among statistical properties, fairness, and computation are studied. We consider how to make accurate and socially desirable decisions. An impossibility theorem on fair Bayesian estimators is shown, and positive results on the satisfaction of many desirable properties are discussed. The chapter also briefly discusses an automated framework for designing decision mechanisms with desirable statistical and fairness properties.

The major goal of this book is to stimulate future research and applications by surveying recent progress in learning and decision-making from rank data as well as to provide a reference to classical definitions, properties, and algorithms for handling rank data.

Lirong Xia
January 2019

Acknowledgments

I would like to thank my coauthors Hossein Azari Soufiani, William Chen, Yiling Chen, Vincent Conitzer, David Hughes, Kevin Hwang, Jeffrey Kephart, Jerome Lang, Haoming Li, Chao Liao, Ao Liu, Zhenming Liu, Pinyan Lu, Nicholas Mattei, David C. Parkes, Peter Piech, Matthew Rognlie, Sujoy Sikdar, Hui Su, Rohit Vaish, Tristan Villamil, Bo Waggoner, Junming Wang, Qiong Wu, Chaonan Ye, and Zhibing Zhao, for great collaborations and inspirations over years on topics covered by or closely related to this book. I would also like to thank Reshef Meir and Shivani Agarwal for extremely useful comments and feedbacks.

I also deeply appreciate the guidance and help from my postdoctoral and graduate school advisors David C. Parkes, Vincent Conitzer, and Mingsheng Ying.

This work is partly supported by NSF #1453542 and ONR #N00014-17-1-2621.

Last but not least, I want to thank my parents, my sister, my wife, and my daughter for giving me a wonderful life.

Lirong Xia
January 2019

CHAPTER 1

Introduction

The field of AI is shifting toward building intelligent systems that can collaborate effectively with people, and that are more generally human-aware.

—One Hundred Year Study on Artificial Intelligence [Stone et al., 2016]

In many situations, intelligent systems collect data of the people and for the people—human preference and behavior data are collected passively or actively, and are used to help people make better decisions. Such data are often represented by *rankings* over a set of alternatives instead of numerical values, and it is important for the intelligent systems to be fast, accurate, and fair. To this end, this book focuses on answering the following three questions.

- Q1: How can we design fast and accurate algorithms for learning from rank data?

- Q2: How can we efficiently elicit preferences from users?

- Q3: How can we make accurate and socially desirable decisions?

Let us look at a few scenarios of learning and decision-making from rank data.

Scenario 1: Social Choice.   Suppose you are voting to make an important social choice, e.g., choosing the next president (see Figure 1.1). Your ballot can be the most-preferred candidate, a ranking over all candidates, or a partial ranking over some candidates. The ballots are then aggregated by a voting rule to decide the winner. As another example, the jury vote to reach a verdict.

What voting rule should be used to make a correct decision while guaranteeing that voters' preferences are handled in a fair way?

Scenario 2: Crowdsourcing.   In crowdsourcing, a complicated task is decomposed into multiple simpler tasks, which are solved by online workers on crowdsourcing platforms. Answers provided by the online workers are aggregated to solve the original task. For example, Pfeiffer et al. [2012] demonstrated a crowdsourcing framework for the following picture ranking task [Horton, 2010]. There are 12 pictures with 318, 335, 342, 344, 355, 381, 383, 399, 422, 447, 460, and 469 nonoverlapping dots, respectively. The goal is to rank the pictures w.r.t. the number of dots in the descending order.

The problem is solved by crowdsourcing on Amazon Mechanical Turk. Each worker is asked to compare a carefully chosen set of pairs of pictures, and answer which one has more

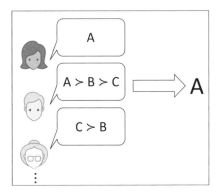

Figure 1.1: Social choice.

dots. For example, Figure 1.2 from Pfeiffer et al. [2012] shows an example of such pairwise comparisons.

In crowdsourcing, how can we compute the most accurate ranking over pictures based on online workers' answers? What are the most informative questions to ask?

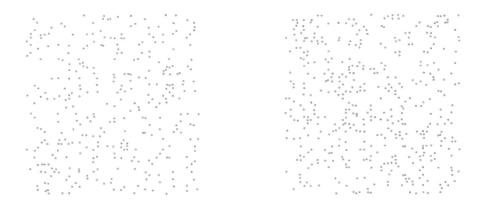

Figure 1.2: The left picture has 342 dots and the right picture has 447 dots [Pfeiffer et al., 2012].

Scenario 3: Group Activity Selection [Darmann et al., 2012].   Suppose you are organizing social activities for some participants. There are three candidate activities: hiking, city tour, and table tennis competition. Different participants may have different preferences over the activities. The participants must be partitioned into two groups, each of which chooses one activity (Figure 1.3).

How should the partition be done based on participants' preferences, and what activity should be chosen for each group?

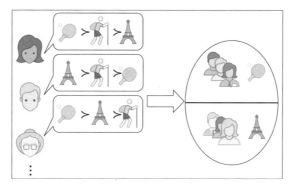

Figure 1.3: Group activity selection.

Scenario 4: Learning to Rank [Liu, 2011].   Suppose you are building a web search engine from a dataset of (keyword, ranking) pairs collected from human annotators, who rank-order the websites according to their relevance to the keywords (Figure 1.4). Each ranking can be a pairwise comparison between webpages (called the *pairwise approach*), or it can be a full ranking over webpages (called the *listwise approach*).

How should the web search engine rank the webpages for a new keyword? If you were to collect new data from human annotators, which keywords would you choose to improve the web search engine?

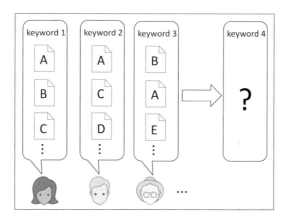

Figure 1.4: Learning to rank.

Scenario 5: Demand Estimation [Train, 2009].   Before the San Francisco Bay Area Rapid Transit (BART) opened in 1975, the Nobel Laureate Daniel McFadden was asked to predict commuters' choice of taking BART over the four existing modes of transportation (1) driving alone, (2) bus + walking to station, (3) bus + driving to station, and (4) carpooling. Seven hundred and seventy-one commuters were surveyed about their top choices from the four existing modes. McFadden's model predicted that 6.3% of commuters would choose BART, which turned out to be very close to the actual share, which was 6.2% (Figure 1.5).

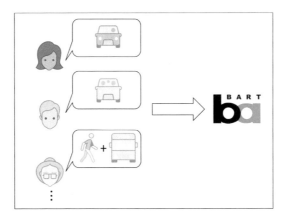

Figure 1.5: Demand estimation.

If more survey data were to be collected, what type of commuters should be targeted and what type of questions should be asked to improve the accuracy of prediction?

Scenario 6: Product Ranking. The Internet Movie Database (IMDB) ranks movies based on online users' ratings, as shown in Figure 1.6. Each user can rate movies from one star to ten stars. Then, a "true Bayesian estimate" is applied to compute the ranking, according to IMDB's official FAQ shown in Figure 1.7. How should the movies, or general products, be ranked in an accurate and fair way?

These scenarios are the tip of an iceberg. Similar problems exist in many other socio-economic systems, such as pricing and product development [Berry et al., 1995], energy forecasting [Goett et al., 2002], meta-search engines [Dwork et al., 2001], recommender systems [Wang et al., 2016], business decision-making [Bhattacharjya and Kephart, 2014], health care [Bockstael, 1999], security [Yang et al., 2011], among many others.

1.1   THE RESEARCH PROBLEM

This book focuses on the problem of learning and decision-making from rank data. As demonstrated in previous scenarios, there is a finite set $\mathcal{A}$ of alternatives and a group of n agents. Each agent uses a ranking R over $\mathcal{A}$ to represent his or her preferences. The collection of agents' preferences, called the *rank data* or *preference profile*, is denoted by $D = (R_1, \ldots, R_n)$, where for each $j \leq n$, R_j is the ranking provided by agent j.

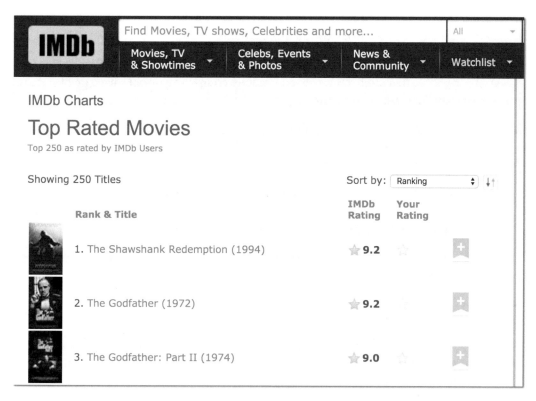

Figure 1.6: IMDB's top 250 movie ranking, July 2018.

How do you calculate the rank of movies and TV shows on the Top Rated Movies and Top Rated TV Show lists?

The following formula is used to calculate the Top Rated 250 titles. This formula provides a true 'Bayesian estimate', which takes into account the number of votes each title has received, minimum votes required to be on the list, and the mean vote for all titles:

weighted rating (WR) = $(v \div (v+m)) \times R + (m \div (v+m)) \times C$

Where:
R = average for the movie (mean) = (rating)
v = number of votes for the movie = (votes)
m = minimum votes required to be listed in the Top Rated list (currently 25,000)
C = the mean vote across the whole report

Figure 1.7: IMDB ranking criterion, July 2018.

Need for New Methodologies.   In the past few decades, technological advancements, especially in internet and computing devices, have brought the following new challenges that cannot be handled efficiently and effectively by existing methods.

1. *Rank Data.* Rank data are becoming more common, as online voting platforms make it easier for people to submit rankings, and mobile computing devices have great potential to learn users' preferences over time.

2. *Interaction.* In addition to passively observing and learning people's preferences, computing devices can actively query peole about their preferences.

3. *Human-aware computer-aided decision-making.* More powerful computers are capable of running more complicated algorithms and models to improve decision-making. Because new technologies can be a double-edged sword that can potentially harm humans, socio-economic criteria need to be considered.

Evaluation Criteria.   To embrace new technologies, it is the time to develop new methodologies for AI-aided decision-making w.r.t. the following criteria (Figure 1.8).

- *Computational criteria* are critical for timely decision-making.

- *Statistical criteria* evaluate the quality of decisions in the statistical sense.

- *Socio-economic criteria* include various desirable normative properties in social choice theory and sociology, such as fairness, strategy-proofness, ethics, etc.

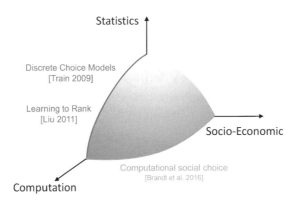

Figure 1.8: Evaluation criteria.

While computationally efficient algorithms for decision-making are always desirable, the importance of statistical criteria and socio-economic criteria depends on the application. For example, in Scenario 1 (social choice), sometimes statistical criteria are more important, e.g., in jury systems, because the goal is to discover the ground truth. On the other hand, in presidential elections, fairness also requires significant consideration. In Scenario 2 (crowdsourcing), Scenario 4 (learning to rank), and Scenario 5 (demand estimation), statistical criteria are arguably more important than fairness, while other socio-economic properties, for example agents' strategic behavior, are important considerations. In Scenario 3 (group activity selection) and Scenario 6 (product ranking), both statistical criteria and fairness are important.

Previous Work.   There is a large body of literature in *discrete choice models* [Train, 2009], where statistical models were built and algorithms were designed to predict agents' probabilities of choosing an alternative from a finite set of alternatives. Theoretical analyses and applications of discrete choice models had been pioneered by the Nobel Laureate Daniel McFadden. Most research in discrete choice models focused on *choice data*, which is a special case of rank data.

In machine learning, various models and methods for preference learning and learning to rank [Liu, 2011] have been developed and widely applied in information retrieval and recommender systems.

In economics, social choice theory focuses on the design and evaluation of group decision-making mechanisms with desirable socio-economic properties. While social choice theory has a long history, modern social choice theory was pioneered by the Nobel Laureate Kenneth Arrow, since his discovery of the celebrated *Arrow's impossibility theorem* [Arrow, 1963]. Computational aspects of social choice have been considered in *computational social choice theory* [Brandt et al., 2016].

1.2   OVERVIEW OF THE BOOK

As mentioned in the beginning of this chapter, this book will discuss some recent progress in learning and decision-making from rank data to answer the following three questions, each of which corresponds to one aspect discussed in the previous section.

Q1: How can we design fast and accurate algorithms for learning from rank data?   The majority part of the book aims at answering this question for widely studied and widely applied statistical models for rank data. After introducing random utility models and distance-based models in Chapter 2, we will discuss several classical and state-of-the-art parameter estimation algorithms in Chapter 3. Chapter 4 covers a general framework for handling rank data called *rank-breaking*. Chapter 5 focuses on the theory and algorithms of mixture models for rank data.

Q2: How can we efficiently elicit preferences from users?   This question will be the main topic of Chapter 6, where we will take a Bayesian experimental design and active learning approach by computing the most cost-effective queries. The chapter will also cover some Bayesian approaches toward learning from partial orders. Preliminary experiments show that choice data are not as cost-effective as other types of rank data.

Q3: How can we make accurate and socially desirable decisions?   Chapter 7 covers a statistical decision-theoretic framework for social choice, under which tradeoffs among statistical properties, fairness, and computation are studied. An impossibility theorem on fair Bayesian estimators will be shown, and positive results on the satisfaction of many desirable properties will also be discussed. The chapter will also briefly discuss an automated framework for designing decision mechanisms with desirable statistical and fairness properties.

CHAPTER 2

Statistical Models for Rank Data

All models are wrong, but some models are useful. —George E.P. Box (1979)

The value of statistical models is that they provide a principled way to make useful decisions. Whether a specific model is useful largely depends on the application domain, and is beyond the scope of this book.

Chapter Overview. This chapter will cover two well studied and widely applied classes of statistical models for rank data: random utility models (Section 2.2) and distance-based models (Section 2.3). For each class of models, we will discuss basic properties and efficient sampling algorithms. Additionally, we will discuss the relationship between random utility models and discrete choice models.

2.1 BASICS OF STATISTICAL MODELING

Let us start with the standard definitions of statistical models for arbitrary type of data. Later in the section we will introduce special statistical models for rank data.

Definition 2.1 Statistical Models. A statistical model $\mathcal{M} = (\mathcal{S}, \Theta, \vec{\pi})$ consists of three components.

1. **Sample space** $\mathcal{S}$ is the set of all possible data (outcomes of experiments).

2. **Parameter space** Θ is the set of indices to probability distributions in $\vec{\pi}$.

3. **Distributions** $\vec{\pi} = \{\pi_{\vec{\theta}} : \vec{\theta} \in \Theta\}$ consist of probability distributions $\pi_{\vec{\theta}}$ over $\mathcal{S}$, one for each parameter $\vec{\theta}$.

The parameter space is not necessary but still included in the definition above, because a learned parameter often has natural explanations that are useful in many applications. The parameter space and the sample space can be discrete or continuous, one-dimensional or multidimensional.

Example 2.2 The set of all Gaussian distributions with fixed variance 1 and unknown means is a statistical model, where the sample space is $\mathbb{R}$, the parameter space is $\mathbb{R}$ (all possible means), and for each $\theta \in \mathbb{R}$, $\pi_\theta(\cdot)$ is the Gaussian distribution whose mean is θ. The set of all Gaussian distributions with unknown variance and unknown mean is also a statistical model, where $\mathcal{S} = \mathbb{R}$, $\Theta = \mathbb{R} \times \mathbb{R}_{>0}$, and each $\vec{\theta} = (\mu, \sigma) \in \Theta$ represents the mean and standard deviation of $\pi_{\vec{\theta}}$.

Definition 2.3   Rank Data.   The following notation will be used in this book.

- **Alternatives** $\mathcal{A}$ is a set of $m \in \mathbb{N}$ objects ranked by the agents.

- **Linear orders** $\mathcal{L}(\mathcal{A})$ is the set of all antisymmetric, transitive, and total binary relations over $\mathcal{A}$. For any linear order $R \in \mathcal{L}(\mathcal{A})$ and any pair of different alternatives $a, b \in \mathcal{A}$, we write $a \succ_R b$ if a is preferred to b in R.

- **Preference profile**, or **rank data** $D = (R_1, \ldots, R_n) \in \mathcal{L}(\mathcal{A})^n$ represent the preference data collected from n agents, where for each $j \leq n$, the j-th agent uses a linear order R_j to represent his or her preferences.

Alternatives are often denoted by lowercase letters, e.g., $\mathcal{A} = \{a, b, c, \ldots\}$, or sometimes by a lowercase letter indexed by $1, \ldots, m$, e.g., $\mathcal{A} = \{a_1, \ldots, a_m\}$. Throughout the book, we will use m to denote the number of alternatives. Clearly, $|\mathcal{L}(\mathcal{A})| = m!$. A linear order is also called a *total order* and is often written as $[a \succ b \succ \cdots]$. For example,

$$\mathcal{L}(\{a, b, c\}) = \left\{ \begin{array}{lll} [a \succ b \succ c], & [a \succ c \succ b], & [b \succ a \succ c] \\ [b \succ c \succ a], & [c \succ a \succ b], & [c \succ b \succ a] \end{array} \right\}.$$

Throughout the book, we will use n to denote the number of agents. When the agents are anonymized, a preference profile can be written as a multi-set in the following way:

$$D = \{n_1 @ T_1, n_2 @ T_2, \ldots, n_k @ T_k\},$$

where $n_1, \ldots, n_k \in \mathbb{N}$ such that $\sum_{i=1}^{k} n_i = n$, and $\{T_1, \ldots, T_k\}$ are k different linear orders. This means that D consists of n_1 copies of T_1, n_2 copies of T_2, etc.

Example 2.4   Let $m = 3$ and $\mathcal{A} = \{a, b, c\}$. Let $D = ([a \succ b \succ c], [c \succ b \succ a], [a \succ b \succ c])$, where the first agent and the third agent rank a at the top, b in the middle, and c in the last place, and the second agent's preferences are the opposite. After anonymization,

$$D = \{2 @ [a \succ b \succ c], 1 @ [c \succ b \succ a]\}.$$

For any $i \leq m$, let $R[i] \in \mathcal{A}$ denote the alternative that is ranked at the i-th position in R. For any $a \in \mathcal{A}$, let $R^{-1}[a] \leq m$ denote the rank of a in R. It follows that $a \succ_R b$ if and only if $R^{-1}[a] < R^{-1}[b]$.

We will view a preference profile as an outcome of a statistical model, where each ranking is generated i.i.d. given the parameter (sometimes called the *ground truth parameter*). Therefore, in the statistical model it suffices to specify the distribution over a single linear order.

Definition 2.5 Statistical Model for Rank Data. Given a set $\mathcal{A}$ of m alternatives and n agents, in a *statistical model for rank data* $\mathcal{M} = (\mathcal{S}, \Theta, \vec{\pi})$, we have:

- $\mathcal{S} = \mathcal{L}(\mathcal{A})^n$; and

- for each $\vec{\theta} \in \Theta$, $\pi_{\vec{\theta}}$ is a distribution over $\mathcal{L}(\mathcal{A})$ such that for any $D \in \mathcal{L}(\mathcal{A})^n$, $\pi_{\vec{\theta}}(D) = \prod_{R \in D} \pi_{\vec{\theta}}(R)$.

One desirable property for statistical models is *identifiability*, which requires that different parameters lead to different distributions over the sample space.

Definition 2.6 Identifiability. A statistical model $\mathcal{M} = (\mathcal{S}, \Theta, \vec{\pi})$ is *identifiable*, if for any pair of different parameters $\vec{\theta}, \vec{\theta}' \in \Theta$, $\pi_{\vec{\theta}} \neq \pi_{\vec{\theta}'}$.

For example, both statistical models in Example 2.2 are identifiable. If a model is not identifiable, then there exist two different parameters that are indistinguishable from data. This can be critical for applications that require explanations of the parameter, for example, interpreting the parameter as the population mean. If the goal is to make decisions just based on the distribution $\pi_{\vec{\theta}}$, then identifiability is not necessary.

Given an identifiable model, a good parameter estimation algorithm $f : \mathcal{S} \to \Theta$ should be able to reveal the ground truth parameter $\vec{\theta}^*$ with high probability, when the data are generated from the model given $\vec{\theta}^*$ and the data size is large. This desirable property is called *consistency*, defined as follows.

Definition 2.7 Consistency. Given a statistical model for rank data $\mathcal{M} = (\mathcal{S}, \Theta, \vec{\pi})$, a parameter estimation algorithm f is *consistent*, if for any $\vec{\theta}^* \in \Theta$, we have:

$$\lim_{n \to \infty} f(D_n) \xrightarrow{p} \vec{\theta}^*,$$

where D_n is a preference profile of n linear orders randomly generated from $\pi_{\vec{\theta}^*}$.

Consistency only concerns whether $f(D_n)$ converges to the ground truth parameter as $n \to \infty$. The rate of convergence is captured by *sample complexity* [Mitzenmacher and Upfal, 2017, Chapter 14], which will not be discussed in this book.

2.1.1 MODELING PARTIAL ORDERS AS EVENTS

In many situations, agents' preferences can only be represented by *partial orders*. For example, in social choice (Scenario 1 in Chapter 1), a voter may not know enough about some presidential candidates to form preferences over them. Most part of this book focuses on linear orders because it provides the basis for handling general partial orders. We will see later in Sections 2.2.4 and 4.2.2 some generic ways to deal with partial-order data.

Definition 2.8 Partial Orders. Given a set of alternatives $\mathcal{A}$, let $\mathrm{PO}(\mathcal{A})$ denote the set of antisymmetric and transitive binary relations over $\mathcal{A}$. For any $Q \in \mathrm{PO}(\mathcal{A})$ and any pair of different alternatives $a, b \in \mathcal{A}$, we write $a \succ_Q b$ if a is preferred to b in Q. For any Q, let $\mathrm{Ext}(Q) \subseteq \mathcal{L}(\mathcal{A})$ denote the set of all linear *extensions* of Q, that is,

$$\mathrm{Ext}(Q) = \{R \in \mathcal{L}(\mathcal{A}) : \forall a, b \in \mathcal{A}, a \succ_Q b \Rightarrow a \succ_R b\}.$$

It follows that $\mathcal{L}(\mathcal{A}) \subseteq \mathrm{PO}(\mathcal{A})$. That is, linear orders are special cases of partial orders. A partial order can be visualized as a directed acyclic graph as in the following example.

Example 2.9 A partial order Q over $\mathcal{A}$ can be visualized as a directed acyclic graph $G = (\mathcal{A}, E)$ where there is an edge $a \to b$ in E if and only if $a \succ_Q b$. For example, a linear order $[a \succ b \succ c]$ is shown in Figure 2.1a. A partial order and its linear extensions are shown in Figure 2.1b.

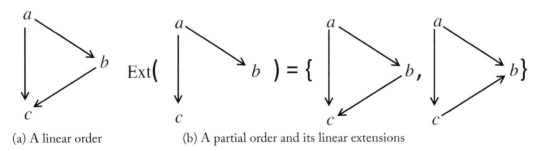

(a) A linear order   (b) A partial order and its linear extensions

Figure 2.1: Graphical representation of partial orders and extensions.

One way to deal with partial orders is to build new statistical models, where the sample space consists of partial orders. This approach can be useful in some cases, by relating the visibility of an alternative to the ground truth parameter. In this book, we will take a different, yet generic approach, by treating partial orders as marginal events in existing statistical models for linear orders.

Definition 2.10 Partial Orders as Events. For any partial order $Q \in \mathrm{PO}(\mathcal{A})$, let $[Q] = \mathrm{Ext}(Q)$ denote the event that consists of linear extensions of Q.

Given a statistical model for rank data, a parameter $\vec{\theta}$, and a partial order Q, let $\pi_{\vec{\theta}}(Q)$ denote the probability of event $[Q]$. In other words,

$$\pi_{\vec{\theta}}(Q) = \pi_{\vec{\theta}}(\text{Ext}(Q)) = \sum_{R \in \text{Ext}(Q)} \pi_{\vec{\theta}}(R).$$

For example, a commonly studied special case of partial orders is pairwise comparisons. For any pair of different alternatives $a, b \in \mathcal{A}$, let $[a \succ b] = \{R \in \mathcal{L}(\mathcal{A}) : a \succ_R b\}$ denote the event "a is preferred to b."

2.2   RANDOM UTILITY MODELS

In the picture ranking task (Scenario 2 in Chapter 1), an online worker's behavior can be modeled in the following way: the worker first estimates the numbers of dots in both pictures, then chooses the picture with the larger estimated number. The estimated number of dots in a picture can be seen as the ground truth number plus a random noise.

This is the rationale behind the random utility models (RUMs). An RUM assumes that each alternative a_i is characterized by an unknown distribution π_i. Each π_i is chosen from a set of distributions $\vec{\pi}_i$ indexed by Θ_i. To generate a ranking under an RUM, an agent first samples her latent value u_i for each alternative a_i from π_i, then ranks the alternatives w.r.t. their latent values in descending order. Following the convention, we will call such latent values "latent utilities."[1] For example, the process of generating $[a_2 \succ a_1 \succ a_3]$ is illustrated in Figure 2.2.

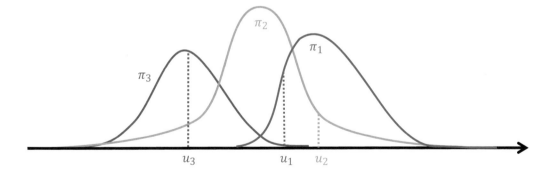

Figure 2.2: Generating $a_2 \succ a_1 \succ a_3$.

Formally, the parameterized family of distributions for each alternative is denoted by a statistical model $\mathcal{M}_i = (\mathbb{R}, \Theta_i, \vec{\pi}_i)$, where the sample space is $\mathbb{R}$ (the latent utility); Θ_i is the

---

[1]This does not mean that the agents have utilities over the alternatives as in the utility theory.

parameter space, where a typical parameter is denoted by $\vec{\theta}_i$; and $\vec{\pi}_i = \{\pi_{\vec{\theta}_i} : \vec{\theta}_i \in \Theta_i\}$ are the distributions over $\mathbb{R}$.

Definition 2.11   Random Utility Model (RUM).   Given a set of m alternatives $\mathcal{A}$, a set of m statistical models $\{\mathcal{M}_i = (\mathbb{R}, \Theta_i, \vec{\pi}_i) : i \leq m\}$, and the number of rankings n, let $\mathrm{RUM}(\mathcal{M}_1, \ldots, \mathcal{M}_m)$ denote the *random utility model* over $\mathcal{A}$, where:

- the sample space is $\mathcal{L}(\mathcal{A})^n$;

- the parameter space is $\Theta = \Theta_1 \times \cdots \times \Theta_m$; and

- for any $\vec{\theta} = (\vec{\theta}_1, \ldots, \vec{\theta}_m) \in \Theta_1 \times \cdots \times \Theta_m$ and any linear order $R = [a_{i_1} \succ a_{i_2} \succ \ldots \succ a_{i_m}]$,

$$
\pi_{\vec{\theta}}(R) = \int_{-\infty}^{\infty} \pi_{i_m, \vec{\theta}_{i_m}}(u_{i_m}) \int_{u_{i_m}}^{\infty} \pi_{i_{m-1}, \vec{\theta}_{i_{m-1}}}(u_{i_{m-1}}) \cdots \int_{u_{i_2}}^{\infty} \pi_{i_1, \vec{\theta}_{i_1}}(u_{i_1}) du_{i_1} \cdots du_{i_m}.
$$

(2.1)

Equivalently, $\pi_{\vec{\theta}}(R)$ can be computed by the integration from the top alternative a_{i_1}:

$$
\pi_{\vec{\theta}}(R) = \int_{-\infty}^{\infty} \pi_{i_1, \vec{\theta}_{i_1}}(u_{i_1}) \int_{-\infty}^{u_{i_1}} \pi_{i_2 \vec{\theta}_{i_2}}(u_{i_2}) \cdots \int_{-\infty}^{u_{i_{m-1}}} \pi_{i_m, \vec{\theta}_{i_m}}(u_{i_m}) du_{i_m} \cdots du_{i_1}.
$$

(2.2)

Ties among alternatives are not considered in Definition 2.11 because under natural conditions, in particular when the probability density function (PDF) of each utility distribution is continuous, the probability for two latent utilities to be the same is 0.

Example 2.12   Thurstone's Case V Model [Thurstone, 1927].   For any $i \leq m$, let $\mathcal{M}_i$ be the set of Gaussian distributions with fixed variance and unknown mean, as in Example 2.2. $\mathrm{RUM}(\mathcal{M}_1, \ldots, \mathcal{M}_m)$ is known as Thurstone's Case V model for rank data.

RUMs with location families is a useful subclass of RUMs, where each $\mathcal{M}_i$ is a set of distributions parameterized by their means. In other words, within each $\mathcal{M}_i$, the shapes of distributions are the same.

Definition 2.13   RUM with Location Families.   A *location family* is a statistical model $(\mathbb{R}, \mathbb{R}, \vec{\pi})$, where there exists a distribution π^0 whose mean is 0, such that for any $\theta \in \mathbb{R}$, $\pi_\theta(x) = \pi^0(x - \theta)$. $\mathrm{RUM}(\mathcal{M}_1, \ldots, \mathcal{M}_m)$ is called an *RUM with location families*, if for all $i \leq m$, $\mathcal{M}_i$ is a location family.

For example, Thurstone's Case V model in Example 2.12 is an RUM with location families.

2.2.1   THE PLACKETT–LUCE MODEL

One key challenge in applying RUMs is that no closed-form formula for the probability of rankings in (2.1) or (2.2) is known in general. Therefore, statistical inference for general RUMs, such as computing the maximum likelihood estimators, is often computationally hard.

The only known exception is the Plackett–Luce model. In this section, we will introduce two equivalent definitions of the Plackett–Luce model with different parameterizations. The first definition, which is called the $\vec{\gamma}$ *parameterization*, assumes that each alternative is characterized by a positive number that represents its likelihood of being ranked higher than other alternatives.

Definition 2.14   Plackett–Luce Model, The $\vec{\gamma}$ Parameterization.   Given a set of m alternatives $\mathcal{A}$ and a number n, the Plackett–Luce model is a statistical model for rank data, where

- the parameter space Θ is $\mathbb{R}^m_{>0}$, and

- for any $\vec{\gamma} \in \Theta$ and any linear order $R = [a_{i_1} \succ a_{i_2} \succ \cdots \succ a_{i_m}]$,

$$\pi_{\vec{\gamma}}(R) = \underbrace{\frac{\gamma_{i_1}}{\sum_{l=1}^m \gamma_{i_l}}}_{a_{i_1}} \times \underbrace{\frac{\gamma_{i_2}}{\sum_{l=2}^m \gamma_{i_l}}}_{a_{i_2}} \times \cdots \times \underbrace{\frac{\gamma_{i_{m-1}}}{\gamma_{i_{m-1}} + \gamma_{i_m}}}_{a_{i_m}}. \tag{2.3}$$

The Plackett–Luce model (sometimes called Plackett–Luce or PL for short) has the following intuitive explanation. Suppose there are m balls in a urn. Each alternative a_i is represented by a ball, whose size is γ_i. Then, a linear order is generated in the following m stages. In stage $t = 1$ to m, a remaining ball a_{i_t} is taken out with probability that is proportional to its size γ_{i_t}, among the remaining balls in the urn, which is $\frac{\gamma_{a_{i_t}}}{\sum_{l=t}^m \gamma_{a_{i_l}}}$. The order of drawing corresponds to the linear order over the alternatives.

Example 2.15   Let $\mathcal{A} = \{a, b, c\}$ and $\vec{\gamma} = (0.1, 0.4, 0.5)$ for a, b, c, respectively, in the Plackett–Luce model. Figure 2.3 shows the probability distribution $\pi_{\vec{\gamma}}$.

Plackett–Luce is the RUM with location families, where the statistical model for each alternative consists of Gumbel distributions with unknown means. Recall that the PDF of the Gumbel distribution is $\pi_{\text{Gum}}(x) = e^{-x - e^{-x}}$. This observation leads to the following equivalent definition of Plackett–Luce using the $\vec{\theta}$ parameterization as in RUMs.

Definition 2.16   Plackett–Luce Model, The $\vec{\theta}$ Parameterization.   Let $\mathcal{M}_{\text{Gum}}$ denote the set of Gumbel distributions with unknown means, such that for any θ and any $x \in \mathbb{R}$, $\pi_\theta(x) = \pi_{\text{Gum}}(x - \theta)$.[2] The Plackett–Luce model is $\text{RUM}(\mathcal{M}_{\text{Gum}}, \dots, \mathcal{M}_{\text{Gum}})$.

---

[2]Notice that the mean of π_{Gum} is non-zero. This means that $\mathcal{M}_{\text{Gum}}$ is a location family under a slightly different parameterization.

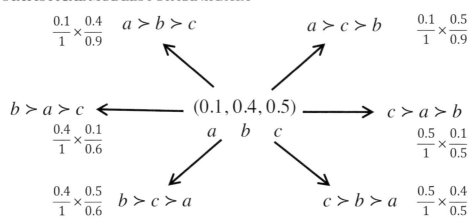

Figure 2.3: $\pi_{(0.1,0.4,0.5)}$ under the Plackett–Luce model.

The following calculation shows the equivalence between the two definitions for Plackett–Luce (Definition 2.16 and Definition 2.14) for $m = 3$ with $\gamma_i = e^{\theta_i}$. The general case can be proved similarly. W.l.o.g. let $R = [a_1 \succ a_2 \succ a_3]$. For any $\vec{\theta} = (\theta_1, \theta_2, \theta_3)$, we have:

$$\pi_{\vec{\theta}}(R) = \int_{-\infty}^{\infty} \pi_{\theta_1}(u_1) \int_{-\infty}^{u_1} \pi_{\theta_2}(u_2) \int_{-\infty}^{u_2} \pi_{\theta_3}(u_3) du_3 du_2 du_1 \tag{2.2}$$

$$= \int_{-\infty}^{\infty} \pi_{\text{Gum}}(u_1 - \theta_1) \int_{-\infty}^{u_1} \pi_{\text{Gum}}(u_2 - \theta_2) \int_{-\infty}^{u_2} \pi_{\text{Gum}}(u_3 - \theta_3) du_3 du_2 du_1$$

$$= \int_{-\infty}^{\infty} \pi_{\text{Gum}}(u_1 - \theta_1) \int_{-\infty}^{u_1} \pi_{\text{Gum}}(u_2 - \theta_2) \cdot e^{-e^{-u_3}} \Big|_{\infty}^{u_2 - \theta_3} du_2 du_1$$

$$= \int_{-\infty}^{\infty} \pi_{\text{Gum}}(u_1 - \theta_1) \int_{-\infty}^{u_1} e^{\theta_2} \cdot e^{-u_2 - e^{-u_2}(e^{\theta_2} + e^{\theta_3})} du_2 du_1 r$$

$$= \int_{-\infty}^{\infty} \pi_{\text{Gum}}(u_1 - \theta_1) \cdot \frac{e^{\theta_2}}{e^{\theta_2} + e^{\theta_3}} \cdot e^{-(e^{\theta_2} + e^{\theta_3})e^{-u_2}} \Big|_{-\infty}^{u_1} du_1$$

$$= \int_{-\infty}^{\infty} \frac{e^{\theta_2}}{e^{\theta_2} + e^{\theta_3}} \cdot e^{\theta_1} \cdot e^{-u_1 - (e^{\theta_1} + e^{\theta_2} + e^{\theta_3})e^{-u_1}} du_1$$

$$= \frac{e^{\theta_2}}{e^{\theta_2} + e^{\theta_3}} \cdot \frac{e^{\theta_1}}{e^{\theta_1} + e^{\theta_2} + e^{\theta_3}} \cdot e^{-(e^{\theta_1} + e^{\theta_2} + e^{\theta_3})e^{-u_1}} \Big|_{-\infty}^{\infty}$$

$$= \frac{e^{\theta_2}}{e^{\theta_2} + e^{\theta_3}} \cdot \frac{e^{\theta_1}}{e^{\theta_1} + e^{\theta_2} + e^{\theta_3}} = \frac{\gamma_1}{\gamma_1 + \gamma_2 + \gamma_3} \cdot \frac{\gamma_2}{\gamma_2 + \gamma_3}. \tag{2.3}$$

2.2.2   PROPERTIES OF RANDOM UTILITY MODELS (RUMS)

The first property says that in any RUM, given the parameter $\vec{\theta}$, events that correspond to non-overlapping partial orders are independent. Two partial orders Q_1 and Q_2 are *non-overlapping*, if there is no alternative that is compared in Q_1 and is also compared in Q_2.

Property 2.17   Independence of Non-Overlapping Partial Orders.   For any RUM, any $\vec{\theta} \in \Theta$, and any pair of non-overlapping partial orders $Q_1, Q_2 \in \mathrm{PO}(\mathcal{A})$,

$$\pi_{\vec{\theta}}(Q_1 \wedge Q_2) = \pi_{\vec{\theta}}(Q_1) \times \pi_{\vec{\theta}}(Q_2). \tag{2.4}$$

Example 2.18   Let $\mathcal{A} = \{a, b, c, d\}$ and $\vec{\gamma} = (0.1, 0.2, 0.3, 0.4)$ for a, b, c, d, respectively, in the Plackett–Luce model. Let $Q_1 = [a \succ b]$, which is a partial order and also represents the event that consists of 12 linear orders, in each of which a is preferred to b. It can be verified that $\pi_{\vec{\gamma}}(a \succ b) = \frac{1}{3}$. Let $Q_2 = [c \succ d]$ be the event that contains 12 linear orders, in each of which c is preferred to d, and $\pi_{\vec{\gamma}}(c \succ d) = \frac{3}{7}$.

Property 2.17 can be applied because Q_1 and Q_2 do not overlap. The following direct calculation of $\pi_{\vec{\gamma}}(a \succ b, c \succ d)$ verifies Property 2.17:

$$
\begin{aligned}
\pi_{\vec{\gamma}}(a \succ b, c \succ d) =\ & \pi_{\vec{\gamma}}(a \succ b \succ c \succ d) + \pi_{\vec{\gamma}}(a \succ c \succ b \succ d) + \pi_{\vec{\gamma}}(c \succ a \succ b \succ d) \\
& + \pi_{\vec{\gamma}}(a \succ c \succ d \succ b) + \pi_{\vec{\gamma}}(c \succ a \succ d \succ b) + \pi_{\vec{\gamma}}(c \succ d \succ a \succ b) \\
=\ & \frac{1}{10} \times \frac{2}{9} \times \frac{3}{7} + \frac{1}{10} \times \frac{3}{9} \times \frac{2}{6} + \frac{3}{10} \times \frac{1}{7} \times \frac{2}{6} \\
& + \frac{1}{10} \times \frac{3}{9} \times \frac{4}{6} + \frac{3}{10} \times \frac{1}{7} \times \frac{4}{6} + \frac{3}{10} \times \frac{4}{7} \times \frac{1}{3} \\
=\ & \frac{1}{7} = \pi_{\vec{\gamma}}(a \succ b) \times \pi_{\vec{\gamma}}(c \succ d).
\end{aligned}
$$

***Proof sketch for Property* 2.17.**   The distribution over linear orders in an RUM given $\vec{\theta}$ can be represented by a Bayesian network as in Figure 2.4. The Bayesian network representation consists of the following two types of random variables.

- For each $a \in \mathcal{A}$, there is a random variable U_a representing the latent utility of a. The probability distribution for U_a is $\pi_{\vec{\theta}_a}(\cdot)$.

- For each partial order Q, there is a binary random variable X_Q whose parents are the variables of the first type, whose corresponding alternatives are involved in Q. For any valuation of Q's parents, denoted by $\vec{u}_{\mathrm{Par}(Q)}$, we have:

$$
\Pr(Q | \vec{u}_{\mathrm{Par}(Q)}) = \begin{cases} 1 & \text{if } Q \text{ is consistent with } \vec{u}_{\mathrm{Par}(Q)} \\ 0 & \text{otherwise} \end{cases},
$$

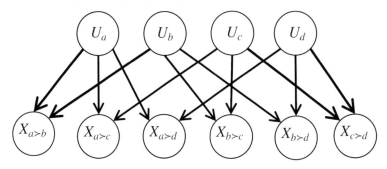

Figure 2.4: The Bayesian network representation of $\pi_{\vec{\theta}}$ in an RUM.

where Q being consistent with $\vec{u}_{\mathrm{Par}(Q)}$ means that Q is consistent with the ordering over the components of $\vec{u}_{\mathrm{Par}(Q)}$.

For example, Figure 2.4 shows the Bayesian network for $\mathcal{A} = \{a, b, c, d\}$. There are four random variables of the first type: $\{U_a, U_b, U_c, U_d\}$, one for each alternative. Among the random variables of the second type, only pairwise comparisons are shown for the purpose of presentation.

Let π_{BN} denote the probability distribution encoded by this Bayesian network. Notice that any assignment of random variables of the second type that receive positive probabilities naturally correspond to a linear order in $\mathcal{L}(\mathcal{A})$, which means that π_{BN} and $\pi_{\vec{\theta}}$ are equivalent in the following sense. For any set of partial orders $Q_1, \ldots Q_k$,

$$\pi_{\vec{\theta}}(Q_1 \wedge \cdots \wedge Q_k) = \pi_{\mathrm{BN}}(Q_1 = \cdots = Q_k = 1).$$

For any pair of non-overlapping Q_1 and Q_2, because their parents are non-overlapping, X_{Q_1} is *d-separated* from X_{Q_1}, which means that X_{Q_1} and X_{Q_2} are independent [Pearl, 1988]. Therefore,

$$\pi_{\vec{\theta}}(Q_1 \wedge Q_2) = \pi_{\mathrm{BN}}(Q_1 = Q_2 = 1)$$
$$= \pi_{\mathrm{BN}}(Q_1 = 1) \times \pi_{\mathrm{BN}}(Q_2 = 1) = \pi_{\vec{\theta}}(Q_1) \times \pi_{\vec{\theta}}(Q_2).$$

This proves Property 2.17.   □

The next property says that the probability of a partial-order event $[Q]$ is the same as that in the sub-model of the RUM with alternatives that are compared in Q.

Property 2.19 Independence of Uncompared Alternatives.   For any RUM$(\mathcal{M}_1, \ldots, \mathcal{M}_m)$ and any partial order Q, let $\mathcal{A}'$ denote the alternatives that are compared in Q and let RUM$(\vec{\mathcal{M}}_{\mathcal{A}'})$ denote the RUM over $\mathcal{A}'$, where the utility distributions for alternative $a_i \in \mathcal{A}'$ is

$\mathcal{M}_i$. For any $\vec{\theta} \in \Theta$, let $\vec{\theta}_{\mathcal{A}'}$ denote its restriction on $\mathcal{A}'$. Then,

$$\underbrace{\pi_{\vec{\theta}}(Q)}_{\text{RUM}(\mathcal{M}_1,\ldots,\mathcal{M}_m)} = \underbrace{\pi_{\vec{\theta}_{\mathcal{A}'}}(Q)}_{\text{RUM}(\vec{\mathcal{M}}_{\mathcal{A}'})}.$$

Example 2.20   Following the setting in Example 2.18, we have that $\mathcal{A}' = \{a, b\}$ is the set of compared alternatives in $Q_1 = [a \succ b]$. $\vec{\gamma}_{\mathcal{A}'} = (0.1, 0.2)$. Therefore, $\pi_{\vec{\gamma}_{\mathcal{A}'}}(a \succ b) = \frac{0.1}{0.1+0.2} = \pi_{\vec{\gamma}}(a \succ b)$.

The following property holds for RUMs with location families, which says that any probability distribution $\pi_{\vec{\theta}}$ over linear orders stays the same after shifting the components in $\vec{\theta}$ by the same value.

Property 2.21   Invariance to Shifts.   For any RUM with location families, any $\vec{\theta} \in \Theta$, any number $s \in \mathbb{R}$, and any $R \in \mathcal{L}(\mathcal{A})$,

$$\pi_{\vec{\theta}}(R) = \pi_{\vec{\theta}+\vec{1}\cdot s}(R).$$

For example, in the $\vec{\theta}$ parameterization of Plackett–Luce (Definition 2.14), parameter shifting (as defined for the $\vec{\theta}$ parameterization) can be seen as multiplying each γ_i by the same value in the $\vec{\gamma}$ parameterization (Definition 2.14), because $\gamma_i = e^{\theta_i}$. For any $s \in \mathbb{R}$ and any $R = [a_{i_1} \succ \cdots a_{i_m}]$, we have:

$$\pi_{\vec{\gamma}}(R) = \prod_{k=1}^{m-1} \frac{\gamma_{i_k}}{\sum_{l=k}^{m} \gamma_{i_l}} = \prod_{k=1}^{m-1} \frac{e^s \cdot \gamma_{i_k}}{\sum_{l=k}^{m} e^s \cdot \gamma_{i_l}} = \pi_{e^s \cdot \vec{\gamma}}(R).$$

Property 2.21 implies that any RUM with location families is non-identifiable, because for any $\vec{\theta}$, its distribution over $\mathcal{L}(\mathcal{A})$ is the same as the distribution for $\vec{\theta} + \vec{1} \cdot s$. In fact, the next property shows that RUMs with location families are identifiable modulo such equivalence.

Definition 2.22   Equivalent Parameters in RUMs with Location Families.   Given an RUM with location families $(\mathcal{M}_1, \ldots, \mathcal{M}_m)$, a pair of parameters $\vec{\theta}, \vec{\mu} \in \Theta$ are *equivalent*, denoted by $\vec{\theta} \approx \vec{\mu}$, if there exists $s \in \mathbb{R}$ such that $\vec{\mu} = \vec{\theta} + \vec{1} \cdot s$.

Property 2.23   Identifiability of RUMs with Location Families Modulo Equivalence.   For any RUM with location families $(\mathcal{M}_1, \ldots, \mathcal{M}_m)$ and any $\vec{\theta}, \vec{\mu} \in \Theta$ such that $\vec{\theta} \not\approx \vec{\mu}$, we have $\pi_{\vec{\theta}} \neq \pi_{\vec{\mu}}$. In particular, if we require that $\theta_m = 0$, then the RUM is identifiable.

In the majority of the book we will require $\theta_m = 0$ in RUMs with location families to guarantee the identifiability of the model. Another commonly used constraint for Plackett–Luce under the $\vec{\gamma}$ parameterization is to require $\vec{\gamma} \cdot \vec{1} = 1$. Under either constraint, only one parameter from each equivalent class is allowed in the constrained RUM.

2.2.3   SAMPLING FROM RANDOM UTILITY MODELS

Efficient sampling from a statistical model for rank data is useful for generating synthetic datasets for experiments. Because each linear order is generated i.i.d., it suffices to have an efficient sampling algorithm for generating one linear order, and then repeat it n times to generate a preference profile.

Sampling a ranking from Plackett–Luce can be done efficiently as in Algorithm 2.1, which is based on its definition with the $\vec{\gamma}$ parameterization (Definition 2.3) and the explanation right after Definition 2.3.

Algorithm 2.1 Efficient Sampling from Plackett–Luce.

Input: A parameter $\vec{\gamma} = (\gamma_1, \ldots, \gamma_m)$ of Plackett–Luce.
Output: A ranking $R \in \mathcal{L}(\mathcal{A})$ from $\pi_{\vec{\gamma}}(\cdot)$ under Plackett–Luce.

1: Let $R = \emptyset$ and $A = \mathcal{A}$.
2: **for** $t = 1$ to m **do**
3:      Choose an alternative a_{i_t} from A with probability proportional to γ_{i_t}.
4:      $R \leftarrow R \succ a_{i_t}$ and $A \leftarrow A \setminus \{a_{i_t}\}$.
5: **end for**
6: **return** R.

Sampling a linear order from a general RUM can be done efficiently if we have an efficient sampler for each utility distribution. In fact, the sampling algorithm (Algorithm 2.2) follows almost directly from the definition of RUMs: we first generate the latent utilities for all alternatives, then rank-order them as the output.

Algorithm 2.2 Efficient Sampling from RUM.

Input: an RUM, a parameter $\vec{\theta} = (\vec{\theta}_1, \ldots, \vec{\theta}_m)$, and efficient samplers for all $\pi_{\vec{\theta}_i}$.
Output: A ranking $R \in \mathcal{L}(\mathcal{A})$ from $\pi_{\vec{\theta}}(\cdot)$ under the RUM.

1: **for** $i = 1$ to m **do**
2:      Sample u_i from $\pi_{\vec{\theta}_i}$.
3: **end for**
4: Rank order $\{u_1, \ldots, u_m\}$ such that $u_{i_1} \geq u_{i_2} \geq \cdots \geq u_{i_m}$.
5: **return** $R = [a_{i_1} \succ a_{i_2} \succ \cdots \succ a_{i_m}]$.

The key challenge is to efficiently sample a value from $\pi_{\vec{\theta}_i}$. Notice that PDFs of many commonly used distributions are either monotonically decreasing (e.g., PDF of the exponential distribution) or unimodal (e.g., PDF of Gaussian and PDF of Gumbel). Therefore, the *ziggurat algorithm* [Marsaglia and Tsang, 2000] can be applied.

The ziggurat algorithm is a rejection sampling algorithm that works for any distribution with monotonically decreasing PDF. For unimodal distributions, the algorithm divides the PDF into two halves at the peak, and then samples the left part and the right part separately. Figure 2.5 shows an example of the ziggurat algorithm for the right half of the Gumbel distribution.

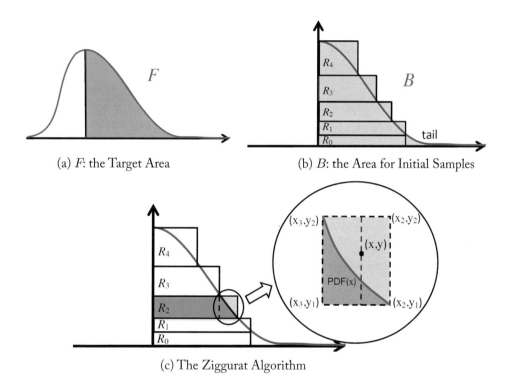

(a) F: the Target Area

(b) B: the Area for Initial Samples

(c) The Ziggurat Algorithm

Figure 2.5: The ziggurat algorithm for sampling from the right half of Gumbel.

It is not hard to see that sampling x from PDF is equivalent to first sampling a point (x, y) uniformly at random from the area beneath the PDF, then output x. For Gumbel, it is the blue area denoted by F in Figure 2.5a. Like other rejection sampling algorithms, the ziggurat algorithm first samples a point (x, y) from an area B that covers the target area F, then tests whether (x, y) indeed belongs to F: if it does, then the sample is accepted and x is returned; otherwise the sample is discarded, and a new sample is generated and tested until acceptance. The closer B is to F, the easier (x, y) is accepted.

For example, the area B covering F is shown in Figure 2.5b in red. In the example, B consists of five rectangles $(R_0, \ldots, R_4)$ and a tail. For each $0 \leq i \leq 4$, let (x_i, y_i) denote the coordinate of the upper right corner of R_i, such that $y_i = \text{PDF}(x_{i+1})$ (assuming $x_5 = 0$).

At this point we already have the following naive rejection sampling algorithm. Let us put R_0 and the tail aside for now and just focus on sampling from $R_1 \cup \cdots \cup R_4$.

- Step 1. Choose a rectangle R_i with probability proportional to its area.

- Step 2. Generate a sample (x, y) inside R_i, which can be done by sampling x and y independently and uniformly.

- Step 3. Check whether $y \leq \text{PDF}(x)$; if so, then return x; otherwise, go back to Step 1.

The ziggurat algorithm has two further improvements.

1. Areas of $R_1, \ldots, R_4$ are the same, which is equivalent to the area of R_0+tail, so that Step 1 chooses i uniformly at random.

2. In Step 3, x is generated first and is accepted if $x \leq x_{i+1}$. Otherwise, y is generated and compared to $\text{PDF}(x)$.

Figure 2.5c shows an example execution of the ziggurat algorithm. Suppose R_2 is selected. If x is no more than x_3, then the sample is automatically accepted; otherwise y is generated to test whether (x, y) is in the blue region or the red region.

For the R_0+tail case, a fallback sampler is used, which can be a different sampling algorithm or another ziggurat algorithm restricted to the tail distribution. The formal ziggurat algorithm is presented in Algorithm 2.3.

2.2.4 CONNECTION TO DISCRETE CHOICE MODELS

Discrete choice models aim at modeling the probability for an agent to choose one alternative when she faces a *choice set* $C \subseteq \mathcal{A}$. Each choice data is a (choice set, choice) pair, denoted by (C, c), where $c \in C$ is the chosen alternative. The model predicts the *choice probabilities*, i.e., the probabilities of choosing each alternative in a choice set $C' \subseteq \mathcal{A}$.

Definition 2.24 Discrete Choice Model. Given a set of alternatives $\mathcal{A}$, a discrete choice model (Θ, Π) consists of two parts:

- Θ is the set of parameters and

- $\Pi = \{\pi_{\vec{\theta},C} : \vec{\theta} \in \Theta, C \subseteq \mathcal{A}\}$. For each parameter $\vec{\theta} \in \Theta$ and each non-empty choice set $C \subseteq \mathcal{A}$, $\pi_{\vec{\theta},C}$ is a probability distribution over C.

Algorithm 2.3 The ziggurat sampling algorithm [Marsaglia and Tsang, 2000].

Input: A monotonically decreasing PDF($\cdot$), pre-computed $R_0, \ldots, R_K$, and a fallback sampler for sampling from R_0+tail.

Output: A sample from PDF($\cdot$).

1: **loop**
2:     Sample $0 \leq i \leq K$ uniformly at random.
3:     **if** $i > 0$ **then**
4:         Sample $x \in [0, x_i]$ uniformly at random.
5:         If $x \leq x_{i+1}$, then **return** x; otherwise sample $y \in [y_{i-1}, y_i]$ uniformly at random, and **return** x only if $y \leq$ PDF(x).
6:     **else**
7:         Sample and **return** x from R_0+tail by the fallback sampler.
8:     **end if**
9: **end loop**

Note that the discrete choice model defined above is not a statistical model. In fact, it is a collection of $2^m - 1$ statistical models, one for each choice set. For example, one of the most widely used discrete choice models, called the *multinomial logit model*, is defined as follows.

Definition 2.25 Multinomial Logit Model. Given a set $\mathcal{A}$ of m alternatives, in the multinomial logit model for discrete choice, $\Theta = \mathbb{R}^m$. For any $\vec{\theta} \in \Theta$, any $C \subseteq \mathcal{A}$, and any $c \in C$,

$$\pi_{\vec{\theta}, C}(c) = \frac{e^{\theta_c}}{\sum_{a \in C} e^{\theta_a}}.$$

Like the Plackett–Luce model, multinomial logit has a random utility interpretation: suppose each alternative a_i is represented by a Gumbel distribution parameterized by its mean θ_i. When an agent faces a choice set C, she first samples a latent utility for each alternative independently, then chooses the one with the highest latent utility. In general, the random utility idea can be used to define discrete choice models as follows.

Definition 2.26 Discrete Choice Model Based on Random Utilities. Given a set of m alternatives $\mathcal{A}$ and a set of m statistical models $\{\mathcal{M}_i = (\mathbb{R}, \Theta_i, \vec{\pi}_i) : i \leq m\}$, let $\mathrm{DCM}(\mathcal{M}_1, \ldots, \mathcal{M}_m)$ denote the discrete choice model, where

- $\Theta = \prod_{i=1}^m \Theta_i$ and

- for any $\vec{\theta} = (\vec{\theta}_1, \ldots, \vec{\theta}_m)$, any choice set $C \subseteq \mathcal{A}$, and any $a_k \in C$, let $\pi_{\vec{\theta}, C}(a_k)$ be the probability that the latent utility for a_k is higher than the latent utilities of alternatives in $C \setminus \{a_k\}$. That is,

$$\pi_{\vec{\theta},C}(a_k) = \underbrace{\int_{-\infty}^{\infty} \pi_{\vec{\theta}_k}(u_k)}_{u_k} \underbrace{\int_{(-\infty,u_k)^{m-1}} \prod_{a_l \in C \setminus \{a_k\}} \pi_{\vec{\theta}_l}(u_l) \, d\vec{u}_{-k} du_k}_{\text{other alternatives in } C}.$$

As another example, the *multinomial probit model* is the discrete choice model where each utility distribution is a Gaussian distribution.

Given a dataset of n discrete choices $\{(C_j, c_j) : j \leq n\}$, a common approach for parameter estimation under a discrete choice model is the maximum likelihood method:

$$\arg\max_{\vec{\theta}} \prod_{j=1}^{n} \pi_{\vec{\theta},C_j}(c_j). \tag{2.5}$$

This should not be confused with the maximum likelihood estimator (MLE), which is typically defined for statistical models.

The remainder of this section reveals a connection between $\mathrm{RUM}(\mathcal{M}_1, \ldots, \mathcal{M}_m)$ and $\mathrm{DCM}(\mathcal{M}_1, \ldots, \mathcal{M}_m)$. We will first see how a parameter estimation algorithm for $\mathrm{DCM}(\mathcal{M}_1, \ldots, \mathcal{M}_m)$, which is based on choice data, can be used to estimate parameters for $\mathrm{RUM}(\mathcal{M}_1, \ldots, \mathcal{M}_m)$. The idea is quite natural: given a dataset of linear orders, we will first convert it to a dataset of choice data, then apply an algorithm for $\mathrm{DCM}(\mathcal{M}_1, \ldots, \mathcal{M}_m)$ to compute an estimate to $\mathrm{RUM}(\mathcal{M}_1, \ldots, \mathcal{M}_m)$.

The conversion leads computationally efficient algorithms for RUM with rank data by leveraging algorithms for discrete choice models, because the likelihood maximization problem (2.5) is often easier to solve than computing the MLE for general RUMs with rank data. This idea will be further explored in Chapter 4.

Applying Discrete Choice Models to Rank Data.   As discussed above, we just need a principled way of converting a linear order to choice data. For the logit model, a linear order $R = [a_{i_1} \succ \cdots \succ a_{i_m}]$ is converted to the following choice data:

$$\forall 1 \leq k \leq m - 1, (\{a_{i_k}, a_{i_{k+1}}, \ldots, a_{i_m}\}, a_{i_k}).$$

The idea is quite natural: a_{i_1} being ranked at the top position of R means that a_{i_1} is chosen when the agent faces all alternatives. Similarly, a_{i_2} should be chosen when the agent faces $\{a_{i_2}, \ldots, a_{i_m}\}$, etc. This is similar to the discussion after Definition 2.14.

Then, the maximum likelihood method (2.5) for multinomial logit becomes the same as the MLE of Plackett–Luce in Definition 2.14. In fact, using multinomial logit to learn from linear-order data is known as *exploded logit* [Train, 2009].

For the probit model and other RUMs, a linear order $[a_{i_1} \succ \cdots \succ a_{i_m}]$ is converted to $\binom{m}{2}$ pairwise comparison data, each of which can be seen as a choice data about two alternatives:

$$\forall k < l, (\{a_{i_k}, a_{i_l}\}, a_{i_k}).$$

Applying RUM for Rank Data to Choice Data.   Notice that choice data are a special case of partial orders. Let us take a quick look at two approaches.[3]

The first approach views each choice data point (C, c) as a partial order $\{c \succ a : a \in C\}$, which is an event in the RUM for rank data (Definition 2.10). Given choice data $D_c = \{(C_j, c_j) : j \le n\}$ and a parameter $\vec{\theta}$, the *composite marginal likelihood* is defined as the multiplication of the probabilities of the marginal events in D_c:

$$CL(\vec{\theta}, D_c) = \prod_{j=1}^{n} \pi_{\vec{\theta}}(C_j, c_j).$$

Then, the parameter of the RUM for rank data can be computed by maximizing the composite marginal likelihood. For example, given parameter $\vec{\theta}$ under the Plackett–Luce model, the probability of the event represented by choice set (C, c) is $\dfrac{e^{\theta_c}}{\sum_{a \in C} e^{\theta_a}}$, according to Property 2.19. $CL(\vec{\theta}, D_c)$ then becomes the maximum likelihood method (2.5) for the logit model. Similar composite marginal likelihood methods will be discussed for parameter estimation under RUMs (Section 4.2.2) and for Bayesian preference elicitation (Section 6.3).

The second approach builds a latent variable model for choice data as shown in Figure 2.6.

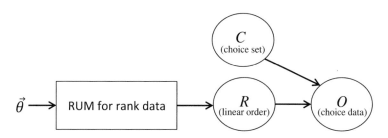

Figure 2.6: Latent variable model for choice data.

In Figure 2.6, given a parameter $\vec{\theta}$, the RUM for rank data generates a linear order $R \in \mathcal{L}(\mathcal{A})$. A choice set C is independently generated. O represents the choice data, and it takes (C, c) with probability 1, where c is the top-ranked alternative in C according to R. This model can be used to learn from choice data and predict discrete choice probabilities.

Take Plackett–Luce again for example. Let C be chosen from $2^{\mathcal{A}} \setminus \{\emptyset\}$ uniformly at random. For any choice data D_c, the MLE of the latent model in Figure 2.6 is the same as the maximum likelihood method (2.5) for the logit model.

---

[3]The general question is how to learn from partial orders under RUM, which is beyond the scope of this book. See Sections 4.2.2 and 6.3 for some theories and algorithms.

2.3 DISTANCE-BASED MODELS

Distance-based models are another widely studied class of statistical models for rank data. In a distance-based model, a distance function $d(\vec{\theta}, R)$ is used to measure the closeness between a parameter $\vec{\theta}$ and a linear order R. The probability of generating R decreases exponentially with its distance to $\vec{\theta}$.

Definition 2.27 Distance-Based Model with Fixed Dispersion. Given a parameter space Θ, a distance function $d : \Theta \times \mathcal{L}(\mathcal{A}) \to \mathbb{R}_{\geq 0}$, and a *dispersion value* $0 < \varphi < 1$, a *distance-based model* $\mathcal{M} = (\mathcal{L}(\mathcal{A})^n, \Theta, \vec{\pi})$ is a statistical model for rank data, where for any $\vec{\theta} \in \Theta$ and any $R \in \mathcal{L}(\mathcal{A})$,

$$\pi_{\vec{\theta}}(R) \propto \varphi^{d(\vec{\theta}, R)}.$$

In this book, we assume that the dispersion value φ is fixed except in Section 5.4. For any $\vec{\theta} \in \Theta$, let $Z_{\vec{\theta}} = \sum_{R \in \mathcal{L}(\mathcal{A})} \varphi^{d(\vec{\theta}, R)}$ denote the normalization factor at $\vec{\theta}$. It follows that $\pi_{\vec{\theta}}(R) = \varphi^{d(\vec{\theta}, R)}/Z_{\vec{\theta}}$.

2.3.1 MALLOWS' MODEL

Mallows' model is a distance-based model, where $\Theta = \mathcal{L}(\mathcal{A})$ and the distance function is the *Kendall–Tau* distance, which is the total number of disagreements in pairwise comparisons between alternatives in the parameter and in a data point, both of which are linear orders.

Definition 2.28 Kendall-Tau Distance. For any pair of linear orders V, W in $\mathcal{L}(\mathcal{A})$, let $KT(V, W)$ denote the *Kendall-Tau distance* between V and W, such that

$$KT(V, W) = |\{a, b\} \subseteq \mathcal{A} : a \succ_V b \text{ and } b \succ_W a|.$$

Example 2.29 $KT([a \succ b \succ c], [c \succ b \succ a]) = 3$. $KT([a \succ b \succ c], [b \succ c \succ a]) = 2$, where pairwise comparisons between $\{a, b\}$ and between $\{a, c\}$ are different; and the pairwise comparisons between $\{b, c\}$ are the same.

Definition 2.30 Mallows' Model with Fixed Dispersion [Mallows, 1957]. Given $0 < \varphi < 1$ and $n \in \mathbb{N}$, *Mallows' model with fixed dispersion* is denoted by $\mathcal{M}_{\text{Ma},\varphi} = (\mathcal{L}(\mathcal{A})^n, \mathcal{L}(\mathcal{A}), \vec{\pi})$, where the parameter space is $\mathcal{L}(\mathcal{A})$ and for any $V, W \in \mathcal{L}(\mathcal{A})$, $\pi_W(V) = \frac{1}{Z_{m,\varphi}} \varphi^{KT(V,W)}$, where $Z_{m,\varphi}$ is the normalization factor.

It is not hard to prove by induction that for Mallows' model:

$$Z_{m,\varphi} = \sum_{V \in \mathcal{L}(A)} \varphi^{KT(V,W)} = \frac{\prod_{i=2}^{m}(1-\varphi^i)}{(1-\varphi)^{m-1}}.$$

Example 2.31 Figure 2.7 shows $\pi_{a \succ b \succ c}$ under Mallows' model.

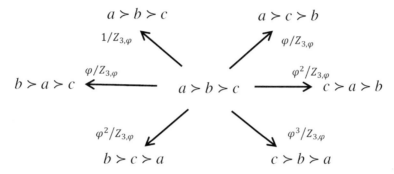

Figure 2.7: $\pi_{a \succ b \succ c}(\cdot)$ under Mallows' model.

For any $k \leq m$, we let $N_k = 1 + \varphi + \cdots + \varphi^{k-1}$. The following property characterizes the probability for a given set of alternatives to be ranked at the top places in a linear order generated from Mallows' model.

Property 2.32 Unranked-Top-k Probability, see e.g., Lemma 2.5 in Awasthi et al. [2014].
Given φ, W, any $k \leq m$, and any $1 \leq i_1 < i_2 < \cdots < i_k$. In Mallows' model with dispersion φ, the probability for $\{W[i_1], W[i_2], \ldots, W[i_k]\}$ to be the top-k alternatives is:

$$\frac{\prod_{l=1}^{k} \sum_{t=0}^{l-1} \varphi^t}{N_{m-k+1} \times \cdots \times N_m \times \varphi^{k(k+1)/2}} \cdot \varphi^{i_1 + \cdots + i_k}.$$

We recall that $W[i]$ is the alternative that is ranked at the i-th position in W. Property 2.32 is called "unranked," because we are not required to rank $\{W[i_1], W[i_2], \ldots, W[i_k]\}$, which are the top-$k$ alternatives. We will use the repeated insertion model introduced in the next subsection to prove this property. Now let us look at an example of Property 2.32.

Example 2.33 By Property 2.32, we have the following probabilities, where $k = 1, 2, 3$, respectively:

- for any $i_1 \leq m$, $\pi_W(W[i_1]$ is ranked at the top$) = \dfrac{\varphi^{i_1 - 1}}{N_m}$; and

- for any $i_1 < i_2 \leq m$, $\pi_W(\{W[i_1], W[i_2]\} \text{ are top 2}) = \dfrac{1 + \varphi}{N_{m-1}N_m\varphi^3} \cdot \varphi^{i_1+i_2}$; and

- for any $i_1 < i_2 < i_3 \leq m$,

$$\pi_W(\{W[i_1], W[i_2], W[i_3]\} \text{ are top 3}) = \frac{(1+\varphi)(1+\varphi+\varphi^2)}{N_{m-2}N_{m-1}N_m\varphi^6} \cdot \varphi^{i_1+i_2+i_3}.$$

2.3.2   REPEATED INSERTION MODEL: EFFICIENT SAMPLING FROM MALLOWS

In this section, we will see an efficient sampling method for Mallows' model as the result of the *repeated insertion model* [Doignon et al., 2004]. Given a parameter $W \in \mathcal{L}(\mathcal{A})$, a linear order R is constructed in m steps. Initially all alternatives are unranked and R is empty. In each step, the top unranked alternative according to W is inserted to R, such that the probability of inserting the alternative to location k in R is proportional to φ^{-k}. This process is formally shown in Algorithm 2.4.

Algorithm 2.4 Efficient Sampling from Mallows [Doignon et al., 2004].

Input: A linear order $W \in \mathcal{L}(\mathcal{A})$ and $0 < \varphi < 1$.
Output: A ranking $R \in \mathcal{L}(\mathcal{A})$ from π_W under Mallows' model.

1: Let $R = \emptyset$.
2: **for** $i = 1$ to m **do**
3:     Insert $W[i]$ to location $1 \leq k \leq i$ in R with probability $\dfrac{\varphi^{i-k}}{1 + \varphi + \cdots + \varphi^{i-1}}$.
4: **end for**
5: **return** R.

Example 2.34   Figure 2.8 shows the three steps of Algorithm 2.4 for $m = 3$ and $W = [a \succ b \succ c]$. In the first step, a is inserted to R. In the second step, b is inserted to R with two possibilities: $b \succ a$ with probability $\frac{\varphi}{1+\varphi}$ and $a \succ b$ with probability $\frac{1}{1+\varphi}$. In the third step, c is inserted to R with probability shown in the figure.

The probability of generating the linear order in a leaf node is the product of probabilities along the path from the root to the leaf. For example, the probability of generating $[a \succ b \succ c]$ is $\frac{1}{1+\varphi} \cdot \frac{1}{1+\varphi+\varphi^2} = \frac{1}{Z_{3,\varphi}}$ and the probability of generating $[b \succ c \succ a]$ is $\frac{\varphi}{1+\varphi} \cdot \frac{\varphi}{1+\varphi+\varphi^2} = \frac{\varphi^2}{Z_{3,\varphi}}$. It can be verified that the probabilities are the same as in Figure 2.7.

We now use the repeated insertion model to prove Property 2.32. W.l.o.g. let $W = [a_1 \succ a_2 \succ \cdots \succ a_m]$. For any $l \leq k$, generating a linear order where $\{a_{i_1}, \ldots, a_{i_k}\}$ (with $i_1 < i_2 <$

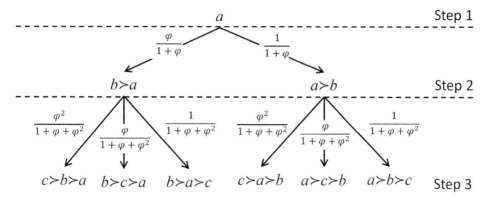

Figure 2.8: Repeated insertion model for $m = 3$.

$\cdots < i_k$) are the top-k alternatives can be viewed as generating a linear order in the repeated insertion model in the following $k + 1$ stages.

- **Stage 1** (Inserting $a_1, \ldots, a_{i_1}$): $a_1, \ldots, a_{i_1-1}$ can be inserted at arbitrary positions. Then, a_{i_1} must be inserted at the top position, which happens with probability $\frac{\varphi^{i_1-1}}{N_{i_1}}$.

- **Stage l for each $2 \leq l \leq k$**: in the first half of the stage, each a_t with $i_{l-1} + 1 \leq t \leq i_l - 1$ is inserted to a position in $[l, t]$, which happens with probability $\frac{N_{t-l}}{N_t}$. Then, a_{i_l} is inserted to a position in $[1, l]$, which happens with probability $\frac{\varphi^{i_l-1} + \cdots + \varphi^{i_l-l}}{N_{i_l}}$.

- **Stage $k + 1$**: for each $t \geq i_k + 1$, a_t is inserted to a position in $[k, t]$, which happens with probability $\frac{N_{t-k}}{N_t}$.

Consequently, the probability for $\{a_{i_1}, \ldots, a_{i_k}\}$ to be the top-k alternatives is:

$$\underbrace{\frac{\varphi^{i_1-1}}{N_{i_1}}}_{\text{Stage 1}} \times \cdots \times \underbrace{\frac{N_{i_{l-1}+1-l}}{N_{i_{l-1}+1}} \times \cdots \frac{N_{i_l-1-l}}{N_{i_l-1}} \times \frac{\varphi^{i_l-1} + \cdots + \varphi^{i_l-l}}{N_{i_l}}}_{\text{Stage } l} \times$$

$$\cdots \times \underbrace{\frac{N_{i_k+1-k}}{N_{i_k+1}} \times \cdots \times \frac{N_{m-k}}{N_m}}_{\text{Stage } k+1}$$

$$= \frac{\prod_{l=1}^{k} \sum_{t=0}^{l-1} \varphi^t}{N_{m-k+1} \times \cdots \times N_m \times \varphi^{k(k+1)/2}} \cdot \varphi^{i_1 + \cdots + i_k}.$$

2.3.3   CONDORCET'S MODEL

Condorcet's model is the distance-based model where the parameter space is the set of all possibly cyclic rankings over $\mathcal{A}$, denoted by $\mathcal{B}(\mathcal{A})$, and the distance function is the Kentall–Tau distance.

More precisely, let $\mathcal{B}(\mathcal{A})$ denote the set of all irreflexive, antisymmetric, and total binary relations over $\mathcal{A}$. It follows that $|\mathcal{B}(\mathcal{A})| = 2^{\binom{m}{2}}$. Any $Q \in \mathcal{B}(\mathcal{A})$ can be visualized as a direct (possibly cyclic) graph $G = \{\mathcal{A}, E\}$, where there is an edge $a \rightarrow b$ in E if and only if $a \succ_Q b$.

Example 2.35   Figure 2.9 shows $\mathcal{B}(\{a, b, c\})$.

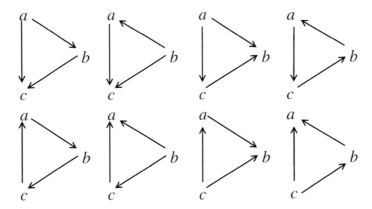

Figure 2.9: $\mathcal{B}(\{a, b, c\})$: irreflexive, antisymmetric, and total binary relations over $\{a, b, c\}$.

It follows that $\mathcal{L}(\mathcal{A}) \subseteq \mathcal{B}(\mathcal{A})$. The Kendall–Tau distance can be naturally extended to $\mathcal{B}(\mathcal{A}) \times \mathcal{L}(\mathcal{A})$ by counting the number of disagreements in pairwise comparisons.

Definition 2.36   Condorcet's Model with Fixed Dispersion [Condorcet, 1785, Young, 1988].   Given $0 < \varphi < 1$, *Condorcet's model with fixed dispersion* is denoted by $\mathcal{M}_{\text{Co},\varphi} = (\mathcal{L}(\mathcal{A})^n, \mathcal{B}(\mathcal{A}), \vec{\pi})$, where the parameter space is $\mathcal{B}(\mathcal{A})$, and for any $W \in \mathcal{B}(\mathcal{A})$ and $V \in \mathcal{L}(\mathcal{A})$, $\pi_W(V) = \frac{1}{Z_{m,\varphi}} \varphi^{\text{KT}(V,W)}$, where $Z_{m,\varphi} = (1 + \varphi)^m$ is the normalization factor.

2.4   DATASETS AND MODEL FITNESS

There are many datasets on rankings and preferences publicly available online, including Preflib [Mattei and Walsh, 2013]; Movielens data [Harper and Konstan, 2015]; label preferences [Hüllermeier et al., 2008]; Microsoft learning to rank data [Qin and Liu, 2013]; and preferences over cars by MTurk users [Abbasnejad et al., 2013]. Google has also launched a dataset search tool.[4]

[4]https://toolbox.google.com/datasetsearch

Unfortunately, no ground truth parameter is available in many real-world preference datasets. Therefore, statistical models are often measured by their fitness to data. Intuitively, a statistical model $\mathcal{M}$ fits a dataset D, if the model contains a distribution $\pi_{\vec{\theta}}$ over the sample space that is "close" to the dataset, which can be measured by $\pi_{\vec{\theta}}(D)$, the probability of generating the dataset under the distribution, also called the *likelihood* of $\vec{\theta}$. Therefore, it is natural to use the maximum likelihood of parameters in $\mathcal{M}$ as a measure of fitness. But only using the maximum likelihood is problematic due to overfitting. Consider the model that contains all distributions over the sample space. Then, for every dataset D, the model always contains a distribution with the maximum likelihood (compared to any parameter in any model).

The following commonly used criteria consider the number of parameters in the model to discourage overfitting. In each criterion, l^* is the maximum likelihood of parameters in the model, d is the dimension of the parameter space in the model, and n is the number of data points.

- **Akaike Information Criterion** [Akaike, 1974]: $\text{AIC} = 2d - 2\ln(l^*)$.

- **Corrected Akaike Information Criterion** [Hurvich and Tsai, 1989]: $\text{AICc} = \text{AIC} + \frac{2d(d+1)}{n-d-1}$.

- **Bayesian Information Criterion** [Schwarz, 1978]: $\text{BIC} = d\ln(n) - 2\ln(l^*)$.

A smaller AIC, AICc, or BIC means better fitness. Table 2.1 from Zhao et al. [2018b] summarizes the percentage of one model to be strictly better than another model w.r.t. the three criteria, for 209 datasets of linear orders on Preflib. k-RUM (k-PL) represents the mixture model of k Gaussian RUM (Plackett–Luce), which will be introduced and discussed in Chapter 5.

Observations.   Table 2.1 shows that the three information criteria agree on the following order over models:

$$k\text{-RUM} \succ k\text{-PL} \succ \text{RUM} \succ \text{PL},$$

where $A \succ B$ means that the number of datasets where A beats B is more than the number of datasets where B beats A. We note that the 40.2% for k-RUM vs. RUM w.r.t. BIC means that k-RUM is strictly better than RUM in 40.2% of the datasets. In other datasets the optimal k is achieved at $k = 1$.

2.5   BIBLIOGRAPHICAL NOTES

Section 2.1.   Definitions of statistical models, events, and identifiability can be found in standard textbooks in statistics.

Section 2.2.   Random utility models were introduced by Thurstone [1927]. The Plackett–Luce model is named after Plackett [1975] and Luce [1977]. The Plackett–Luce model restricted to pairwise comparisons is known as the *Bradley–Terry* model [Bradley and Terry, 1952]. The

Table 2.1: Comparisons of model fitness on Preflib data [Zhao et al., 2018b]. Each row represents a combination of a fitness measure (AIC, AICc, or BIC) and a model (k-RUM, k-PL, RUM, or PL). Each column represents the other model in comparison (k-RUM, k-PL, RUM, or PL). The numbers in the table represent the percentage of datasets where the row model fits better than the column model w.r.t. the criterion of the row. Numbers > 50% are highlighted.

		k-RUM	k-PL	RUM	PL
AIC	k-RUM	0	**60.8%**	**60.3%**	**90.0%**
	k-PL	39.2%	0	**79.4%**	**90.4%**
	RUM	0	20.6%	0	**76.6%**
	PL	10.0%	0	23.4%	0
AICc	k-RUM	0	**60.3%**	**59.8%**	**90.0%**
	k-PL	39.7%	0	**79.4%**	**89.5%**
	RUM	0	20.6%	0	**76.6%**
	PL	10.0%	0	23.4%	0
BIC	k-RUM	0	**66.0%**	40.2%	**84.2%**
	k-PL	34.0%	0	**59.8%**	**66.0%**
	RUM	0	40.2%	0	**76.6%**
	PL	15.8%	0	23.4%	0

Plackett–Luce model restricted to triple-wise comparisons is known as the *Pendergrass–Bradley* model [Pendergrass and Bradley, 1960]. Property 2.17 was proved by Zhao et al. [2016]. Thomas et al. [2007] offers a comprehensive survey of sampling algorithms including the ziggurat algorithm (Algorithm 2.3). The book by Train [2009] offers a comprehensite overview of discrete choice models and algorithms.

Section 2.3.   General distance-based models were introduced by Diaconis, according to Fligner and Verducci [1986]. Properties of Mallows' model are extensively discussed in the book by Marden [1995]. The repeated insertion model was proposed by Doignon et al. [2004] and was generalized by Lu and Boutilier [2014]. Condorcet's model was also studied by Elkind and Shah [2014] and Azari Soufiani et al. [2014b]. There are other commonly studied distance functions, such as Cayley's distance, Spearman's foot rule. Relationships between these measures were discussed by Diaconis and Graham [1977]. Data generation from the distance-based model with Cayley distance can be done efficiently using the procedure by Diaconis and Hanlon [1992].

CHAPTER 3

Parameter Estimation Algorithms

In this chapter, we will focus on parameter estimation algorithms for the statistical models for rank data introduced in Chapter 2. Given a statistical model $\mathcal{M}$ and data D, the goal is to compute parameters that "best" explain the data. One natural approach is to compute the *maximum likelihood estimator (MLE)*:

$$\text{MLE}(D) = \arg\max_{\vec{\theta} \in \Theta} \pi_{\vec{\theta}}(D).$$

Here, $\pi_{\vec{\theta}}(D)$ is also called the likelihood of $\vec{\theta}$, denoted by $\mathcal{L}(\vec{\theta}, D)$. Recall that linear orders in D are i.i.d. Therefore, MLE also maximizes the log-likelihood $\sum_{R \in D} \log \pi_{\vec{\theta}}(R)$.

MLE is favorable in many statistical inference problems not only because it is natural, but also because it satisfies many desirable statistical properties under natural conditions, including *asymptotic efficiency*, which means that no *unbiased* estimator can achieve asymptotically smaller mean squared error, as the data size goes to infinity [Casella and Berger, 2001, Section 10.1.12].

One common drawback of MLE is that it is hard to compute under many statistical models for rank data. In such cases, MLEs are often computed by iterative algorithms. Given limited computational resource, an (iterative) algorithm that aims at computing the MLE may be as accurate as an algorithm that aims at computing a different estimator.

Chapter Overview. Section 3.1 discusses two algorithms for computing the MLE and a generalized method-of-moments (GMM) algorithm for the Plackett–Luce model. Section 3.2 discusses an Expectation-Maximization (EM) algorithm and a GMM algorithm for computing the MLE under general random utility models. Section 3.3 discusses algorithms for computing the MLE under Mallows' model and Condorcet's model.

3.1 ALGORITHMS FOR THE PLACKETT–LUCE MODEL

Recall that the likelihood function under the Plackett–Luce model has a closed-form formula. That is, $\mathcal{L}(\vec{\theta}, D) = \prod_{j=1}^{n} \prod_{i=1}^{m-1} \dfrac{e^{\theta_{R_j[i]}}}{\sum_{k=i}^{m} e^{\theta_{R_j[k]}}}$. Before computing the MLE, we will first examine the uniqueness of MLE solutions. The good news is that for any data D, the likelihood function $\mathcal{L}(\vec{\theta}, D)$ under Plackett–Luce, as a function of the parameters $\vec{\theta}$, is strictly log-concave, as shown in the following theorem by Hunter [2004].

Theorem 3.1 [Hunter, 2004]. *For any preference profile D of linear orders, $\mathcal{L}(\vec{\theta}, D)$ under the Plackett–Luce model is strictly log-concave.*

Theorem 3.1 implies that over any compact parameter space, e.g., $[0, 1]^m$, the maximizer of the likelihood function exists and is unique. However, the uniqueness does not automatically hold for unbounded parameter space, as shown in the following example.

Example 3.2 Let $D = \{[a \succ b \succ c], [a \succ c \succ b]\}$. In Plackett–Luce with $\vec{\theta}$ parameterization (Definition 2.16), let $\theta_c = 0$ to guarantee identifiability (Property 2.23). We have $\mathcal{L}(\vec{\theta}, D) = (\frac{e^{\theta_a}}{e^{\theta_a} + e^{\theta_b} + 1})^2 \cdot \frac{e^{\theta_b}}{e^{\theta_b} + 1} \cdot \frac{1}{e^{\theta_b} + 1}$. Notice that increasing θ_a leads to a strictly higher likelihood, which means that $\mathrm{MLE}(D) = \emptyset$.

Intuitively, the emptiness of $\mathrm{MLE}(D)$ is due to the following phenomenon: the alternatives can be partitioned into two sets A and B, such that there is no evidence in D that any alternative in B is preferred to any alternative in A. Then, by increasing parameters for alternatives in A while keeping ratios among alternatives in A and ratios among alternatives in B the same, the likelihood will increase. This phenomenon can be circumvented by the following condition.

Condition 3.3 Well-Connected Profiles. A preference profile D is *well-connected* if for any partition $\mathcal{A} = A \cup B$ (with $A \cap B = \emptyset$), there exist $a \in A, b \in B$, and $R \in D$ such that $b \succ_R a$.

Equivalently, well-connectedness can be defined as follows. Given any preference profile D, let G_D denote a directed graph, where the nodes are the alternatives and there is an edge $a \to b$ if and only if $a \succ_R b$ for some $R \in D$. Then, D is well-connected if and only if for any pair of different alternatives (a, b), there is a directed path from b to a in G_D.

Example 3.4 $D_1 = \{[a \succ b \succ c], [a \succ c \succ b]\}$ is not well-connected. $D_2 = \{[a \succ b \succ c], [c \succ a \succ b]\}$ is well-connected. G_{D_1} and G_{D_2} are shown in Figure 3.1.

Theorem 3.5 Non-Emptiness of MLE under Plackett–Luce. *Under Plackett–Luce, $\mathrm{MLE}(D) \neq \emptyset$ if and only if D is well-connected.*

Theorems 3.1 and 3.5 together provide the following necessary and sufficient condition for MLE to work under Plackett–Luce. If the preference profile D is well-connected, then the output of MLE is unique; but if the preference profile D is not well-connected, then the output of MLE is empty.

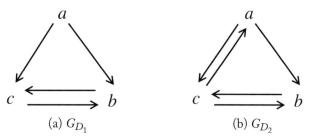

(a) G_{D_1} (b) G_{D_2}

Figure 3.1: G_{D_1} and G_{D_2} for preference profiles in Example 3.4.

Overview of Algorithms. The following subsections describe three parameter estimation algorithms for Plackett–Luce with the $\vec{\gamma}$ parameterization (Definition 2.14).

(1) The Minorize-Maximization (MM) algorithm [Hunter, 2004] in Section 3.1.1 iteratively computes parameters to improve the log-likelihood.

(2) The Luce Spectral Ranking (LSR) algorithm [Maystre and Grossglauser, 2015] in Section 3.1.2 computes the MLE by computing parameters that satisfy the first-order conditions, which suffices because the log-likelihood function is strictly concave.

(3) The generalized method-of-moments (GMM) algorithm [Azari Soufiani et al., 2013a] in Section 3.1.3 does not compute the MLE. Instead, it first establishes *moment conditions* that hold for infinite data generated from the model given the ground truth, then computes an optimal $\vec{\gamma}$ to satisfy the moment conditions given data D.

3.1.1 THE MINORIZE-MAXIMIZATION (MM) ALGORITHM

The Minorize-Maximization (MM)[1] algorithm computes the maximizer of a concave function $f(x)$ by defining a class of *surrogate functions* $\{g_x(y) : x \in \mathbb{R}\}$ such that:

(1) for any x, g_x *minorizes* f, which means that for any y, $g_x(y) \leq f(y)$ and

(2) the equality holds only at $y = x$.

In other words, each surrogate function g_x is below f, and g_x touches f at the unique tangent point x. Figure 3.2 shows a concave function f and three surrogate functions g_{x_1}, g_{x_2}, and g_{x_3}.

Ideally, we would like the maximizer of each surrogate function to be easy to compute and as close to the objective function as possible. Given the surrogate functions, the MM algorithm proceeds in the following steps. First, choose x_1 arbitrarily. Then for each iteration $i = 1, \ldots, T$, let $x_{i+1} = \arg\max_y g_{x_i}(y)$. Finally, the algorithm outputs x_{T+1}. For example, Figure 3.2 shows the first three steps of MM: in the first step, g_{x_1} is used to choose x_2, which maximizes g_{x_1}. Then, x_3 is chosen to maximize g_{x_2} and x_4 is chosen to maximize g_{x_3}.

[1]MM also stands for Majorize-Minimization for miminization problems.

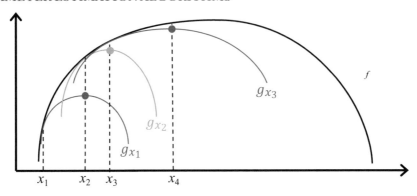

Figure 3.2: The Minorize-Maximization (MM) algorithm.

Hunter's Surrogate Functions for Plackett–Luce.   Hunter [2004] defined the following surrogate functions for the log-likelihood function under Plackett–Luce. For any $\vec{\gamma}, \vec{\gamma}^{(t)} \in (0, \infty)^m$, where $t \in \mathbb{N}$ represents the iteration in Algorithm 3.5 that will be presented soon, let

$$g_{\vec{\gamma}^{(t)}}(\vec{\gamma}) = \sum_{j=1}^{n} \sum_{i=1}^{m-1} \left(\ln \gamma_{R_j[i]} - \frac{\sum_{s=i}^{m} \gamma_{R_j[s]}}{\sum_{s=i}^{m} \gamma_{R_j[s]}^{(t)}} \right) + C. \tag{3.1}$$

In (3.1), C is a constant given $\vec{\gamma}^{(t)}$. We recall that $R_j[s]$ is the alternative that is ranked at the s-th position in R_j. The maximizer $\vec{\gamma}^{(t+1)} = (\gamma_1^{(t+1)}, \ldots, \gamma_{m-1}^{(t+1)})$ of $g_{\vec{\gamma}^{(t)}}(\vec{\gamma})$ can be computed by solving the first-order conditions, leading to the following formulas. Recall that $\gamma_m^{(t+1)}$ is always set to be $e^0 = 1$ for the sake of identifiability (Property 2.23). For each $l \leq m - 1$,

$$\gamma_l^{(t+1)} = \frac{\gamma_l^{(t)}}{\sum_{j=1}^{n} \sum_{i=1}^{m-1} \delta_{ijl} [\sum_{s=i}^{m} \gamma_{R_j[s]}^{(t)}]^{-1}}, \tag{3.2}$$

where δ_{ijl} indicates whether a_l is ranked below a_i in R_j. That is,

$$\delta_{ijl} = \begin{cases} 1 & \text{if } a_l \text{ is ranked below } i \text{ in } R_j \\ 0 & \text{otherwise.} \end{cases} \tag{3.3}$$

The MM algorithm with the surrogate functions (3.1) and their maximizers (3.2) is shown in Algorithm 3.5.

3.1.2   THE LUCE SPECTRAL RANKING (LSR) ALGORITHM

To compute the MLE under Plackett–Luce, the Luce Spectral Ranking (LSR) algorithm [Maystre and Grossglauser, 2015] computes the parameter that satisfies the first-order

---

Algorithm 3.5 The MM Algorithm for Plackett–Luce [Hunter, 2004].

Input: A preference profile D, the number of iterations T.
Output: Estimated parameter of Plackett–Luce.
Initialize Randomly choose $\vec{\gamma}^{(1)}$.

1: **for** $t = 1$ **to** T **do**
2:    Compute $\vec{\gamma}^{(t+1)}$ according to (3.2).
3: **end for**
4: **return** $\vec{\gamma}^{(T+1)}$

---

conditions (FOCs) of the following log-likelihood function:

$$\mathcal{LL}(\vec{\gamma}; D) = \sum_{R \in D} \sum_{i=1}^{m-1} \left(\ln \gamma_{R[i]} - \ln(\sum_{l=i}^{m} \gamma_{R[l]}) \right).$$

Let us use an example to illustrate the idea behind LSR. Let $D = \{2@[a_1 \succ a_2 \succ a_3], 1@[a_3 \succ a_2 \succ a_1]\}$. The first-order condition for γ_1, namely $\frac{\partial \mathcal{LL}(\vec{\gamma}; D)}{\partial \gamma_1} = 0$, is:

$$\text{FOC } \gamma_1 : \ 2(\frac{1}{\gamma_1} - \frac{1}{\gamma_1 + \gamma_2 + \gamma_3}) - \frac{1}{\gamma_1 + \gamma_2 + \gamma_3} - \frac{1}{\gamma_1 + \gamma_2} = 0$$

$$\Longleftrightarrow \frac{1}{\gamma_1 + \gamma_2 + \gamma_3} + \frac{1}{\gamma_1 + \gamma_2} = \frac{2(\gamma_2 + \gamma_3)}{\gamma_1 \cdot (\gamma_1 + \gamma_2 + \gamma_3)}$$

$$\Longleftrightarrow (\frac{1}{\gamma_1 + \gamma_2 + \gamma_3} + \frac{1}{\gamma_1 + \gamma_2}) \cdot \gamma_1 = \frac{2}{\gamma_1 + \gamma_2 + \gamma_3} \cdot \gamma_2 + \frac{2}{\gamma_1 + \gamma_2 + \gamma_3} \cdot \gamma_3.$$

Similarly, the first-order conditions for γ_2 and γ_3 are:

$$\text{FOC } \gamma_2 : \ \frac{3}{\gamma_1 + \gamma_2 + \gamma_3} \cdot \gamma_2 = \frac{1}{\gamma_1 + \gamma_2} \cdot \gamma_1 + \frac{2}{\gamma_2 + \gamma_3} \cdot \gamma_3$$

$$\text{FOC } \gamma_3 : \ (\frac{2}{\gamma_1 + \gamma_2 + \gamma_3} + \frac{2}{\gamma_2 + \gamma_3}) \cdot \gamma_3 = \frac{1}{\gamma_1 + \gamma_2 + \gamma_3} \cdot \gamma_1 + \frac{1}{\gamma_1 + \gamma_2 + \gamma_3} \cdot \gamma_2.$$

Combining the first-order conditions, we have:

$$(\gamma_1, \gamma_2, \gamma_3) \times \underbrace{\begin{bmatrix} -\frac{1}{\gamma_1+\gamma_2+\gamma_3} - \frac{1}{\gamma_1+\gamma_2} & \frac{1}{\gamma_1+\gamma_2} & \frac{1}{\gamma_1+\gamma_2+\gamma_3} \\ \frac{2}{\gamma_1+\gamma_2+\gamma_3} & -\frac{3}{\gamma_1+\gamma_2+\gamma_3} & \frac{1}{\gamma_1+\gamma_2+\gamma_3} \\ \frac{2}{\gamma_1+\gamma_2+\gamma_3} & \frac{2}{\gamma_2+\gamma_3} & -\frac{2}{\gamma_1+\gamma_2+\gamma_3} - \frac{2}{\gamma_2+\gamma_3} \end{bmatrix}}_{M(D,\vec{\gamma})} = 0.$$

This observation can be generalized to arbitrary $m \geq 3$ as follows. Given a preference profile D, let

$$f_{a_l \succ a_i}(\vec{\gamma}) = \sum_{j=1}^{n} \delta_{lji} \frac{1}{\gamma_{jl\downarrow}},$$

where $\gamma_{jl\downarrow} = \sum_{s=R_j^{-1}[a_l]}^{m} \gamma_{R[s]}$, and recall that $R_j^{-1}[a_l]$ is the rank of a_l in R_j, $R_j[s]$ is the alternative ranked at the s-th position in R_j, and δ_{lji} was defined in (3.3). In other words, $\gamma_{jl\downarrow}$ is the sum of γ values of all alternatives that are ranked below a_l in R_j, including a_l itself.

Definition 3.6   Let $M(D, \vec{\gamma})$ denote the matrix where for any $i \neq l$, the (i, l) entry is $f_{a_l \succ a_i}(\vec{\gamma})$; for any $i \leq m$, the (i, i) entry is $-\sum_{l \neq i} f_{a_l \succ a_i}(\vec{\gamma})$.

LSR computes $\vec{\gamma}$ s.t. $\vec{\gamma} \times M(D, \vec{\gamma}) = 0$ by first randomly choosing $\vec{\gamma}^{(1)}$ (e.g., $\vec{1} \cdot \frac{1}{m}$), then iteratively refining $\vec{\gamma}^{(t+1)}$ as the solution to $\vec{\gamma} \times \left(\sum_{l=1}^{t} M(D, \vec{\gamma}^{(l)})\right) = 0$ for $t = 1$ to T.

Algorithm 3.6 The LSR Algorithm for Plackett–Luce [Maystre and Grossglauser, 2015].

Input: A preference profile D and the number of iterations T.
Output: Estimated parameter for Plackett–Luce.
Initialize $\vec{\gamma}^{(1)} = \vec{1} \cdot \frac{1}{m}$, $M^{(1)} = (0)_{m \times m}$.

1: **for** $t = 1$ **to** T **do**
2:     $M^{(t+1)} \leftarrow M^{(t)} + M(D, \vec{\gamma}^{(t)})$.
3:     Compute $\vec{\gamma}^{(t+1)} > \vec{0}$ s.t. $\vec{\gamma}^{(t+1)} \times M^{(t+1)} = 0$.
4: **end for**
5: **return** $\vec{\gamma}^{(T+1)}$.

3.1.3   GENERALIZED METHOD-OF-MOMENTS (GMM) ALGORITHM

Generalized Method-of-Moments (GMM)[2] [Hansen, 1982] is a general class of algorithms for parameter estimation, which has been widely applied, especially when MLE is hard to compute.

High-Level Overview.   The GMM algorithm for Plackett–Luce first defines a column-vector-valued function $\vec{g}(\vec{\gamma}, R)$, called the *moment conditions*, such that $\mathbb{E}_{R \sim \pi_{\vec{\gamma}}}(\vec{g}(\vec{\gamma}, R)) = \vec{0}$, where R is generated from the Plackett–Luce model given parameter $\vec{\gamma}$. Given a preference profile D, the algorithm solves the equations where the expectation is replaced by its empirical counterpart, namely $\sum_{R \in D} \vec{g}(\vec{\gamma}, R) = \vec{0}$.

Let us illustrate the algorithm by an example. Suppose $m = 3$ and the ground truth parameter is $\vec{\gamma} = (\gamma_a, \gamma_b, \gamma_c)$. We have $\Pr_{\vec{\gamma}}(a \succ b) = \frac{\gamma_a}{\gamma_a + \gamma_b}$ and $\Pr_{\vec{\gamma}}(b \succ a) = \frac{\gamma_b}{\gamma_a + \gamma_b}$, which means that

$$\pi_{\vec{\gamma}}(b \succ a) \cdot \gamma_a = \pi_{\vec{\gamma}}(a \succ b) \cdot \gamma_b. \tag{3.4}$$

Similarly, we have $\pi_{\vec{\gamma}}(c \succ a) \cdot \gamma_a = \pi_{\vec{\gamma}}(a \succ c) \cdot \gamma_c$. Therefore,

$$\text{for } \gamma_a : (\pi_{\vec{\gamma}}(b \succ a) + \pi_{\vec{\gamma}}(c \succ a)) \cdot \gamma_a = \pi_{\vec{\gamma}}(a \succ b) \cdot \gamma_b + \pi_{\vec{\gamma}}(a \succ c) \cdot \gamma_c.$$

---

[2]This acronym should not be confused with Gaussian Mixture Models.

Similar equations can be obtained for γ_b and γ_c. Altogether, we have:

$$(\gamma_a, \gamma_b, \gamma_c) \times \underbrace{\begin{bmatrix} -\pi_{\vec{\gamma}}(b \succ a) - \pi_{\vec{\gamma}}(c \succ a) & \pi_{\vec{\gamma}}(b \succ a) & \pi_{\vec{\gamma}}(c \succ a) \\ \pi_{\vec{\gamma}}(a \succ b) & -\pi_{\vec{\gamma}}(a \succ b) - \pi_{\vec{\gamma}}(c \succ b) & \pi_{\vec{\gamma}}(c \succ b) \\ \pi_{\vec{\gamma}}(a \succ c) & \pi_{\vec{\gamma}}(b \succ c) & -\pi_{\vec{\gamma}}(a \succ c) - \pi_{\vec{\gamma}}(b \succ c) \end{bmatrix}}_{P(\vec{\gamma})} = 0.$$

$P(\vec{\gamma})$ will be approximated by its empirical counterpart that is calculated from data D. Let $D[a \succ a']$ denote the number of times $a \succ a'$ in D, it follows from the Law of Large Numbers that

$$P(\vec{\gamma}) = \lim_{n \to \infty} \underbrace{\begin{bmatrix} -D[b \succ a] - D[c \succ a] & D[b \succ a] & D[c \succ a] \\ D[a \succ b] & -D[a \succ b] - D[c \succ b] & D[c \succ b] \\ D[a \succ c] & D[b \succ c] & -D[a \succ c] - D[b \succ c] \end{bmatrix}}_{P(D)} / n . \tag{3.5}$$

Intuitively, when n is large, the solution of $\vec{\gamma} \times P(D) = \vec{0}$ is close to the solution of $\vec{\gamma} \times P(\vec{\gamma}) = \vec{0}$. This relationship holds for all $m \geq 3$ with the following definition of $P(D)$.

Definition 3.7   For any preference profile D over $m \geq 3$ alternatives, let $P(D)$ denote the $m \times m$ matrix, where for any $i \neq t$, the (i, t) entry of $P(D)$ is $D[t \succ i]/n$, and for any $i \leq m$, the (i, i) entry of $P(D)$ is $-\sum_{t \neq i} D[t \succ i]/n$.

We are now ready to present GMM$_{\text{PL}}$ in Algorithm 3.7.

Algorithm 3.7 GMM$_{\text{PL}}$ for Plackett–Luce [Azari Soufiani et al., 2013a] .

Input: A preference profile D.
Output: A set of estimated parameters of Plackett–Luce.
  1: Compute positive solutions $\vec{\Gamma}$ of $\vec{\gamma} \times P(D) = 0$ according to Definition 3.7.
  2: **return** $\vec{\Gamma}$.

Corollary 4.8 in the next chapter shows that the output of Algorithm 3.7 is unique if and only if D is well-connected. We note that GMM$_{\text{PL}}$ does not compute the MLE of Plackett–Luce. Nevertheless, GMM$_{\text{PL}}$ is consistent for Plackett–Luce as we will see in Theorem 4.9.

Now, let us formally introduce the generalized method-of-moments framework for statistical estimation, which will be used later in this book.

Formal Definition of Generalized Method-of-Moments (GMM).   A GMM algorithm for a statistical model $(\mathcal{L}(\mathcal{A})^n, \Theta, \vec{\pi})$ for rank data has two components:

1. a column-vector-valued function $\vec{g} : \mathcal{DS} \times \Theta \to \mathbb{R}^q$ called the *moment conditions*, where $q \in \mathbb{N}$ and each component of $\vec{g}$ is called a moment condition; and

2. an infinite series of $q \times q$ matrices $\mathcal{W} = \{W_t : t \geq 1\}$ that converge to a positive semidefinite matrix W^*.

Note that the moment conditions must be defined as a function of a single data point R and a parameter $\vec{\theta}$, and q may not be the same as m.

Definition 3.8   Generalized Method-of-Moments.   Given the moment conditions $\vec{g}$, $\mathcal{W} = \{W_t : t \geq 1\}$, data D, and $T \in \mathbb{N}$, the GMM is defined as follows:

$$\text{GMM}_T(D) = \{\vec{\theta}^* \in \Theta : \|\vec{g}(D, \vec{\theta}^*)\|_{W_T} = \inf_{\vec{\theta} \in \Theta} \|\vec{g}(D, \vec{\theta})\|_{W_T}\}, \tag{3.6}$$

where $\vec{g}(D, \vec{\theta}) = \frac{1}{n} \sum_{R \in D} \vec{g}(R, \vec{\theta})$ and $\|\vec{a}\|_W = (\vec{a})^T W \vec{a}$.

Therefore, designing a GMM algorithm largely amounts to defining the moment conditions $\vec{g}$ and the matrices $\mathcal{W}$.

Example 3.9   For example, GMM$_{\text{PL}}$ (Algorithm 3.7) for Plackett–Luce with $\vec{\gamma}$ parameterization is a GMM, where:

- the moment conditions are: for any $R \in \mathcal{L}(\mathcal{A})$, let $\vec{g}(R, \vec{\gamma}) = \vec{\gamma} \times P(R)$, where $P(\cdot)$ is the matrix defined in Definition 3.7; and

- $\mathcal{W} = \{W_t = I : t \geq 1\}$.

There are m moment conditions, each of which corresponds to a column in $P(\cdot)$. Recall that $P(D) = \sum_{R \in D} P(R)/n$ and GMM$_{\text{PL}}$ minimizes the L_2 norm of $\vec{\gamma} \times P(D)$. This is equivalent to computing $\vec{\gamma}$ such that $\vec{\gamma} \times P(D) = 0$, as we will see in Theorem 4.5.

Because Θ may not be compact (as in Plackett–Luce with the $\vec{\gamma}$ parameterization in Example 3.9), GMM$_T(D)$ can be empty. It has been proved that under natural conditions GMM satisfies consistency and asymptotic normality, and the optimal W that minimizes the asymptotic variance of the GMM has a closed-form solution [Hansen, 1982]. One critical necessary condition is *global identification*, which states that for any $\vec{\theta}^* \in \Theta$, when the data is generated from $\vec{\theta}^*$, the expected moment conditions hold if and only if $\vec{\theta} = \vec{\theta}^*$.

Condition 3.10   Global Identification   For any parameter $\vec{\theta}^*$, $W_* \times \mathbb{E}_{R \sim \pi_{\vec{\theta}^*}}[\vec{g}(R, \vec{\theta})] = \vec{0}$ if and only if $\vec{\theta} = \vec{\theta}^*$.

3.2   ALGORITHMS FOR GENERAL RANDOM UTILITY MODELS

The major challenge in parameter estimation for general RUM is the lack of closed-form formula for the likelihood function. Moreover, the gradient of (log-)likelihood function is often hard to compute.

In this section, we will introduce efficient algorithms for two classes of RUMs: an EM algorithm for RUMs with *exponential families* in Section 3.2.1, and a GMM algorithm for RUMs with location families (Definition 2.13) in Section 3.2.4.

Definition 3.11 Exponential Family.   A set of distribution $\pi_{\vec{\theta}}$ over $\mathbb{R}$ that are parameterized by $\vec{\theta}$ is called an *exponential family*, if their PDFs have the following format:

$$\forall x \in \mathbb{R}, \pi_{\vec{\theta}}(x) = \exp\left\{\eta(\vec{\theta}) \times T(x) - A(\vec{\theta}) + B(x)\right\}, \tag{3.7}$$

where $A(\vec{\theta})$ and $B(x)$ are scalars, $\eta(\vec{\theta})$ is a row vector and $T(x)$ is a column vector that represents the sufficient statistics of x.

The next example shows that some commonly studied sets of distributions are exponential families.

Example 3.12 Common Distributions as Exponential Families.

- Gaussian distributions with fixed variance σ and unknown mean μ are an exponential family, where there is a single parameter $\vec{\theta} = (\mu)$, $\eta(\mu) = \frac{\mu}{\sigma}$, $T(x) = \frac{x}{\sigma}$, $A(\mu) = \frac{\mu^2}{2\sigma^2}$, and $B(x) = -\frac{x^2}{2\sigma^2} \log \frac{1}{\sqrt{2\pi}\sigma}$.

- Gaussian distributions with unknown variance σ and unknown mean μ are an exponential family, where $\vec{\theta} = (\mu, \sigma)$, $\eta(\mu, \sigma) = (\frac{\mu}{\sigma^2}, -\frac{1}{2\sigma^2})$, $T(x) = (x, x^2)^T$, $A(\mu, \sigma) = \frac{\mu^2}{2\sigma^2} + \ln|\sigma|$, and $B(x) = \ln\frac{1}{\sqrt{2\pi}}$.

- Gumbel distributions with unknown mean μ are an exponential family, where $\vec{\theta} = (\mu)$, $\eta(\mu) = e^{\mu}$, $T(x) = -e^{-x}$, $B(x) = -x$, and $A(\mu) = -\mu$.

Definition 3.13 RUMs with Exponential Families. $RUM(\mathcal{M}_1, \ldots, \mathcal{M}_m)$ is called an *RUM with exponential families*, if for each $i \le m$, $\mathcal{M}_i$ is an exponential family.

3.2.1 THE EXPECTATION-MAXIMIZATION (EM) ALGORITHM

The EM algorithm [Dempster et al., 1977] is a generic and widely applied method for parameter estimation with unobserved data, which computes the parameter with the maximum marginal likelihood of observed data.

Definition 3.14   A statistical model with unobserved data is denoted by $((\mathbf{X}, \mathbf{Z}), \Theta, \vec{\pi})$, where $\mathbf{X}$ and $\mathbf{Z}$ are random variables that denote the observed and unobserved data, respectively; Θ is the parameter space; and $\vec{\pi}$ is a set of distributions over $(\mathbf{X}, \mathbf{Z})$, one for each parameter. The unobserved variable $\mathbf{Z}$ is also called a *latent variable*.

RUM as A Model with Missing Data.   An RUM can be viewed as a statistical model with unobserved data in the following way. The observed data $\mathbf{X} = (R_1, \ldots, R_n)$ are the rankings and the latent variables $\mathbf{Z} = (\vec{u}_1, \ldots, \vec{u}_n)$ are agents' latent utilities for all alternatives, where for any $j \leq n$, $\vec{u}_j = (u_{j1}, \ldots, u_{jm})$ is agent j's latent utilities for all alternatives. For any parameter $\vec{\theta} = (\vec{\theta}_1, \ldots, \vec{\theta}_m)$, any linear order R, and any $\vec{u}_j$, we have:

$$\pi_{\vec{\theta}}(R, \vec{u}_j) = \begin{cases} \prod_{i=1}^{m} \pi_{\vec{\theta}_i}(u_{ji}) & \text{if } R \text{ is consistent with } \vec{u}_j \\ 0 & \text{otherwise.} \end{cases}$$

The EM algorithm computes the parameter $\vec{\theta}$ that maximizes the marginal likelihood of observed data $\mathbf{X} = D$ as follows:

$$\pi_{\vec{\theta}}(D) = \int \pi_{\vec{\theta}}(D, \mathbf{Z}) d\mathbf{Z}.$$

In each iteration t of EM, given the parameter $\vec{\theta}^{(t)}$ computed in the previous iteration, the algorithm consists of an E-step and an M-step.

- **E-step:** Given any $\vec{\theta} = (\vec{\theta}_1, \ldots, \vec{\theta}_m)$, let $Q(\vec{\theta}, \vec{\theta}^{(t)})$ denote the conditional expectation of the complete-data log-likelihood, where the latent variable $\mathbf{Z} = (\vec{u}_1, \ldots, \vec{u}_n)$ is distributed conditioned on $\mathbf{X} = D = (R_1, \ldots, R_n)$ according to $\vec{\theta}^{(t)}$:

$$Q(\vec{\theta}, \vec{\theta}^{(t)}) = \mathbb{E}_{\mathbf{Z}} \left\{ \log \prod_{j=1}^{n} \pi_{\vec{\theta}}(R_j, \vec{u}_j) \mid D, \vec{\theta}^{(t)} \right\}. \tag{3.8}$$

- **M-step:** $\vec{\theta}^{(t+1)}$ is computed as the maximizer of $Q(\vec{\theta}, \vec{\theta}^{(t)})$, that is,

$$\vec{\theta}^{(t+1)} \in \arg\max_{\vec{\theta}} Q(\vec{\theta}, \vec{\theta}^{(t)}).$$

In the remainder of this section we will see an EM algorithm that is tailored for RUMs with exponential families.

3.2.2   EM FOR RUMS: MONTE CARLO E-STEP BY GIBBS SAMPLING

Suppose in an RUM with exponential families, alternative a_i is characterized by the exponential family whose PDF is $\pi_{\vec{\theta}_i}(x) = \exp\left\{\eta_i(\vec{\theta}_i) \times T_i(x) - A_i(\vec{\theta}) + B_i(x)\right\}$. For any $t \in \mathbb{N}$, let $\vec{\theta}^{(t)} = (\vec{\theta}_1^{(t)}, \ldots, \vec{\theta}_m^{(t)})$, where $\vec{\theta}_i^{(t)}$ is the estimated parameter for alternative a_i computed in round $t - 1$.

Then, the E-step (3.8) can be simplified as follows:

$$
\begin{aligned}
Q(\vec{\theta}, \vec{\theta}^{(t)}) &= \mathbb{E}_{\vec{u}} \left\{ \log \prod_{j=1}^{n} \pi_{\vec{\theta}}(R_j, \vec{u}_j) \,\Big|\, D, \vec{\theta}^{(t)} \right\} = \mathbb{E}_{\vec{u}} \left\{ \log \prod_{j=1}^{n} \pi_{\vec{\theta}}(\vec{u}_j) \,\Big|\, D, \vec{\theta}^{(t)} \right\} \\
&= \sum_{j=1}^{n} \sum_{i=1}^{m} \mathbb{E}_{u_{ji}} \left\{ \log \pi_{\vec{\theta}_i}(u_{ji}) \mid R_j, \vec{\theta}^{(t)} \right\} \\
&= \sum_{i=1}^{m} \sum_{j=1}^{n} \left(\eta_i(\vec{\theta}_i) \mathbb{E}_{u_{ji}} \left\{ T_i(u_{ji}) \mid R_j, \vec{\theta}^{(t)} \right\} - A_i(\vec{\theta}_i) \right) + W \\
&= \sum_{i=1}^{m} Q_i(\vec{\theta}_i, \vec{\theta}^{(t)}) + F_t,
\end{aligned}
\tag{3.9}
$$

where $Q_i(\vec{\theta}_i, \vec{\theta}^{(t)}) = \eta_i(\vec{\theta}_i) \left(\sum_{j=1}^{n} \mathbb{E}_{u_{ji}} \left\{ T_i(u_{ji}) \mid R_j, \vec{\theta}^{(t)} \right\} \right) - n A_i(\vec{\theta}_i)$ and

$$
F_t = \sum_{i=1}^{m} \sum_{j=1}^{n} \mathbb{E}_{u_{ji}} \left\{ B_i(u_{ji}) \mid R_j, \vec{\theta}^{(t)} \right\}.
$$

Notice that F_t only depends on $\vec{\theta}^{(t)}$ and D (but does not depend on $\vec{\theta}$). This means that it can be treated as a constant in the M-step. In light of this, it suffices to compute the optimizer for each $Q_i(\vec{\theta}_i, \vec{\theta}^{(t)})$ separately in the M-step, which improves the computational efficiency.

To compute the optimizer for each $Q_i(\vec{\theta}_i, \vec{\theta}^{(t)})$, let $S_j^{i,t+1} = \mathbb{E}_{u_{ji}}\{T_i(u_{ji}) \mid R_j, \vec{\theta}^{(t)}\}$. In other words, $S_j^{i,t+1}$ is the expected sufficient statistics w.r.t. alternative a_i's distribution, when agent j's latent utilities $\vec{u}_j$ is generated from $\pi_{\vec{\theta}^{(t)}}$ conditioned on $\vec{u}_j$ being consistent with R_j.

Because analytical solutions for $S_j^{i,t+1}$ are generally unknown, a Monte Carlo approximation can be used to estimate $S_j^{i,t+1}$, which involves sampling $\vec{u}_j$ according to $\Pr(\vec{u}_j \mid R_j, \vec{\theta}^{(t)})$ by a Gibbs sampler. Then, $S_j^{i,t+1}$ is approximated by $\frac{1}{N} \sum_{l=1}^{N} T_i(u_{ji}^l)$ where N is the number of samples in the Gibbs sampler and u_{ji}^l is the l-th sample for a_i.

Gibbs Sampling for Computing $S_j^{i,t+1}$ (Algorithm 3.8). The key challenge in sampling $\vec{u}_j$ is to maintain the consistency between the ordering of its component and R_j. This is achieved by Gibbs sampling, where in each round, the sufficient statistic T_i is calculated by sampling the latent utility of one randomly chosen alternative a_i. Suppose that a_i is ranked at position k in R_j. After resampling, the new latent utility of alternative a_i is between the utility of the alternative that is ranked right below a_i, namely at position $k + 1$ in R_j, and the alternative that is ranked at position $k - 1$ in R_j.

Effectively, this is be done by sampling from the truncated distribution for alternative a_i as shown in Figure 3.3a and Algorithm 3.8.

Algorithm 3.8 Gibbs Sampling for $S_j^{i,t+1}$ [Azari Soufiani et al., 2012].

Input: $R_j \in \mathcal{L}(\mathcal{A})$, $\vec{\theta}^{(t)}$, $\vec{u}_j$, $i \leq m$.
Output: One sample for $S_j^{i,t+1}$.

1: Let k denote the rank of alternative a_i in R_j. Let $\text{Trunc}_{\vec{\theta}_i^{(t)}}$ denote the distribution $\pi_{\vec{\theta}_i^{(t)}}(\cdot)$ restricted to $[u_{jR_j[k+1]}, u_{jR_j[k-1]}]$, as shown in Figure 3.3.

2: Sample u from $\text{Trunc}_{\vec{\theta}_i}$.

3: **return** $T_i(u)$.

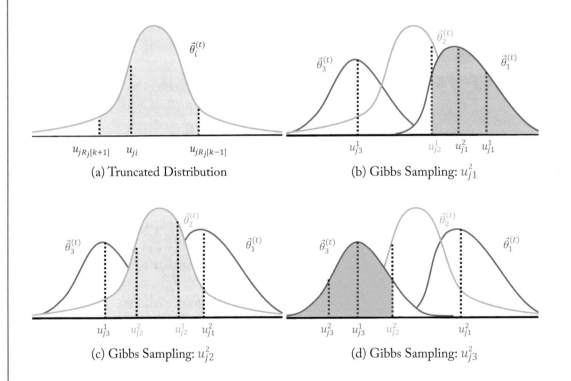

Figure 3.3: Truncated distributions and Gibbs sampling.

Example 3.15   Suppose $R_j = [a_1 \succ a_2 \succ a_3]$. Figure 3.3b,c,d illustrate three calls to the Gibbs sampling algorithm (Algorithm 3.8), where $i = 1, 2, 3$, respectively.

- First call (Figure 3.3b), u_{j1}^2 is sampled to replace u_{j1}^1 from the truncated distribution in blue.

- Second call (Figure 3.3c), u_{j2}^2 is sampled to replace u_{j2}^1 from the truncated distribution in green.

- Third call (Figure 3.3c), u_{j3}^2 is sampled to replace u_{j3}^1 from the truncated distribution in red.

3.2.3   EM FOR RUMS: M-STEP

As (3.9) shows, in the M-step it suffices to compute each $\vec{\theta}_i^{(t+1)}$ separately to maximize

$$Q_i(\vec{\theta}_i, \vec{\theta}^{(t)}) = \eta_i(\vec{\theta}_i) \left(\sum_{j=1}^{n} S_j^{i,t+1} \right) - n A_i(\vec{\theta}_i), \tag{3.10}$$

where $S_j^{i,t+1}$ can be approximately computed by Gibbs sampling as in Algorithm 3.8. We note that these are already much easier to compute than directly optimizing $Q(\vec{\theta}, \vec{\theta}^{(t)})$. Sometimes $\vec{\theta}_i^{(t+1)}$ has a closed-form solution, as shown in the following example.

Example 3.16   Suppose a_i follows a Gaussian distribution with fixed variance σ and unknown mean μ. We recall from Example 3.12 that $\eta(\mu) = \frac{\mu}{\sigma}$ and $A(\mu) = \frac{\mu^2}{2\sigma^2}$. Therefore, $\vec{\theta}_i^{(t+1)} = \frac{\sigma}{n} \sum_{j=1}^{n} S_j^{i,t+1}$. Note that we still need to estimate $S_j^{i,t+1}$, for example by Gibbs sampling.

Putting the E-step and the M-step together, the EM algorithm for RUMs with exponential families is shown in Algorithm 3.9.

Algorithm 3.9 EM algorithm for RUMs with exponential families [Azari Soufiani et al., 2012].

Input: A preference profile D, an RUM with exponential families, T.
Output: Estimated parameter of the RUM.

1: Randomly initialize $\vec{\theta}^{(1)}$.
2: **for** $t = 1$ to T **do**
3:     **E-Step:** For each $i \leq m$ and each $j \leq n$, estimate $S_j^{i,t+1}$ by calling Gibbs Sampling (Algorithm 3.8) multiple times.
4:     **M-Step:** For each $i \leq m$, compute $\vec{\theta}_i^{(t+1)} = \arg\max_{\vec{\theta}_i} Q_i(\vec{\theta}_i, \vec{\theta}^{(t)})$ as in (3.10).
5: **end for**
6: **return** $\vec{\theta}^{(T+1)}$.

3.2.4   GMM FOR RUMS WITH LOCATION FAMILIES

In this section, we will introduce a GMM algorithm for RUMs with location families [Azari Soufiani et al., 2014a]. Recall that in a location family, each utility distribution is parameterized by its mean. Therefore, we will use $\vec{\theta} = (\theta_1, \ldots, \theta_m)$ to denote the parameters, where θ_i is the mean of the utility distribution for alternative a_i. Also recall that θ_m is always set to be 0 for identifiability. For each alternative a_i, let π_i^0 denote its utility distribution with zero mean. It follows that given θ_i, the utility distribution of a_i is $\pi_i^0(x - \theta_i)$.

Let us start with a sufficient condition for the log-likelihood function of an RUM with location families to be concave.

Theorem 3.17 Log-Concavity of Likelihood Function for RUMs with Location Families [Azari Soufiani et al., 2012]. *For any RUM with location families, if the probability density function of each utility distribution is log-concave, then the log-likelihood function is concave.*

Recall from Definition 3.8 that a GMM algorithm consists of a set of moment conditions and a series of weighting matrices $\mathcal{W} = \{W_t : t \geq 1\}$. In this section, we will fix W_t to be the identity matrix and use the moment conditions on probabilities of pairwise comparisons that are similar to the moment conditions in GMM$_{RUM}$ (Algorithm 3.7).

Let us take a closer look at the probabilities of pairwise comparisons in RUM with location families. For any pair of alternatives $a \neq b$, we have:

$$
\begin{aligned}
\pi_{\vec{\theta}}(a \succ b) &= \int_{-\infty}^{\infty} \pi_b^0(u_b - \theta_b) \int_{u_b}^{\infty} \pi_a^0(u_a - \theta_a) du_a \, du_b \\
&= \int_{-\infty}^{\infty} \pi_b^0(x_b) \int_{x_b+\theta_b}^{\infty} \pi_a^0(u_a - \theta_a) du_a \, dx_b && \text{(let } x_b = u_b - \theta_b) \\
&= \int_{-\infty}^{\infty} \pi_b^0(x_b) \int_{x_b+\theta_b-\theta_a}^{\infty} \pi_a^0(x_a) dx_a \, dx_a && \text{(let } x_a = u_a - \theta_a).
\end{aligned}
$$

Therefore, $\pi_{\vec{\theta}}(a \succ b)$ can be written as a function of $\theta_a - \theta_b$, denoted by $f^{ab}(\theta_a - \theta_b)$. More precisely,

$$
f^{ab}(x) = \int_{-\infty}^{\infty} \pi_b^0(x_b) \int_{x_b-x}^{\infty} \pi_a^0(x_a) dx_a \, dx_b. \tag{3.11}
$$

In other words, $f^{ab}(x)$ is the probability that a is preferred to b when the difference between the mean of the distribution for a is x ahead of the mean of the distribution for b.

By definition, we have $f^{ab}(\theta_a - \theta_b) + f^{ba}(\theta_b - \theta_a) = 1$. Next, we define the moment conditions that will be used in the GMM$_{RUM}$ algorithm (Algorithm 3.10).

Definition 3.18   For any ranking $R \in \mathcal{L}(\mathcal{A})$ and any $a, b \in \mathcal{A}$, we let:

$$
X^{a \succ b}(R) = \begin{cases} 1 & \text{if } a \succ_R b \\ 0 & \text{otherwise.} \end{cases}
$$

For each pair of different alternatives a, b, there is a moment condition

$$
g^{ab}(R, \vec{\theta}) = X^{a \succ b}(R) \times f^{ba}(\theta_b - \theta_a) - X^{b \succ a}(R) \times f^{ab}(\theta_a - \theta_b). \tag{3.12}
$$

By definition, $\mathbb{E}_{R \sim \pi_{\vec{\theta}}}[X^{a \succ b}(R)] = f^{ab}(\theta_a - \theta_b)$. We recall from Definition 3.8 that GMM computes $\vec{\theta}$ to minimize $\|\vec{g}(D, \vec{\theta})\|_2^2 = \sum_{a \neq b} g^{ab}(D, \vec{\theta})^2$. This leads to Algorithm 3.10.

Algorithm 3.10 $\text{GMM}_{\text{RUM}}(D)$ [Azari Soufiani et al., 2014a] .

Input: A preference profile D, an RUM with location families.
Output: Estimated parameter of the RUM.

1: For all pairs of alternatives $a \neq b$, compute $X^{a \succ b}(D)$.
2: Compute $\vec{\theta}^* \in \arg\min_{\vec{\theta}} \sum_{a \neq b} (g^{ab}(D, \vec{\theta}))^2$.
3: **return** $\vec{\theta}^*$.

Step 2 in Algorithm 3.10 can use any optimization algorithm, for example gradient descent algorithms. Computing the gradient in Algorithm 3.10 is easier than computing the gradient for the likelihood function, because each g^{ab} only involves two variables, whose gradient involves f^{ab} and $(f^{ab})'$. f^{ab} can be estimated numerically, and $(f^{ab})'$ can be calculated as in the following proposition.

Proposition 3.19 [**Azari Soufiani et al., 2014a**] *For any RUM with location families, where each utility distribution π_i^0 has support $(-\infty, \infty)$, for any different pair of alternatives a and b, f^{ab} is monotonic increasing on $(-\infty, \infty)$ with $\lim_{x \to -\infty} f^{ab}(x) = 0$ and $\lim_{x \to \infty} f^{ab}(x) = 1$. Moreover, if π_a^0 and π_b^0 are continuous, then f^{ab} is continuously differentiable, and*

$$f^{ab}(x)' = \int_{-\infty}^{\infty} \pi_b^0(y)\pi_a^0(y - x)dy.$$

In some cases $(f^{ab})'$ has a closed-form solution, as shown in the following example.

Example 3.20 Consider the RUM with Gaussian distributions whose variances are 1. For any $a \neq b$ we have:

$$f^{ab}(x)' = \frac{1}{2\pi}\int_{-\infty}^{\infty} e^{-\frac{y^2}{2}} e^{-\frac{(y-x)^2}{2}} dy = \frac{1}{2\sqrt{\pi}}e^{-\frac{x^2}{4}}.$$

A similar formula exists when the variances are not the same.

One theoretical guarantee of GMM_{RUM} is its consistency for a large class of RUMs with location families.

Theorem 3.21   Consistency of GMM_{RUM} [Azari Soufiani et al., 2014a]. *For any RUM with location families, if the PDF of each utility distribution π_i^0 is continuous, then GMM_{RUM} (Algorithm 3.10) is consistent.*

As a corollary, GMM_{RUM} for the Plackett–Luce model and GMM_{RUM} for Thurstone's case V model (Example 2.12) are consistent.

3.3   ALGORITHMS FOR DISTANCE-BASED MODELS

In this section, we will introduce algorithms for computing the MLE under distance-based models. Let us start with Mallows' model by revealing a natural equivalence among three problems: (1) MLE under Mallows, (2) the Kemeny voting rule, and (3) a well-known NP-hard problem called *feedback arc set* [Karp, 1972].

Given a preference profile D, the Kemeny voting rule chooses linear orders with minimum Kendall-Tau distance to D. In a FEEDBACK ARC SET problem, we are given a weighted graph G, and we are asked to remove a set of edges with minimum total weight to make the remaining graph acyclic.

Definition 3.22   Weighted Majority Graph.   For any preference profile D, its *weighted majority graph*, denoted by $\text{WMG}(D)$, is a complete weighted directed graph $(\mathcal{A}, E)$ with weights w_D, where

- the vertices are the alternatives $\mathcal{A}$, and

- the weight on any edge $a \to b$ is the winning margin of a over b in their head-to-head competition. That is, for any $a, b \in \mathcal{A}$,

$$w_D(a \to b) = \#\{R \in D : a \succ_R b\} - \#\{R \in D : b \succ_R a\}.$$

Example 3.23   Figure 3.4 shows a preference profile and its weighted majority graph. Only edges with strictly positive weights are shown.

Ranking R_1 and R_2:	$a \succ c \succ d \succ b$
Ranking R_3 and R_4:	$b \succ c \succ d \succ a$
Ranking R_5 and R_6:	$d \succ a \succ b \succ c$
Ranking R_7:	$b \succ c \succ a \succ d$
Ranking R_8:	$d \succ b \succ a \succ c$

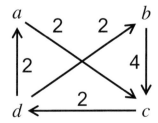

(a) Preference profile D of 8 rankings.   (b) Weighted majority graph WMG (D).

Figure 3.4: A preference profile D and its weighted majority graph.

In the weighted majority graph, $w_D(a \to c) = 2$ because $a \succ c$ in five rankings $(R_1, R_2, R_5, R_6, R_8)$ and $c \succ a$ in two rankings (R_3, R_4, R_7); $w_D(a \to b) = w_D(b \to a) = 0$ because $a \succ b$ in four rankings (R_1, R_2, R_5, R_6) and $b \succ a$ in three rankings (R_3, R_4, R_7, R_8).

Theorem 3.24   *MLE for Mallows computes the outcome of the Kemeny rule, which is the set of all linearizations of the remaining weighted majority graph after removing edges in a minimum feedback arc set.*

Proof. The equivalence between the MLE under Mallows and the outcome of Kemeny follows after the definition. To see that any linear order in the outcome of Kemeny corresponds to a linearization of a remaining WMG, for any linear order W, we write the Kendall-Tau distance between W and D as follows:

$$\text{KT}(W, D) = \sum_{a \succ_W b} \#\{R \in D : b \succ_R a\} = \sum_{a \succ_W b} (n + w_D(b \to a))/2$$
$$= nm(m - 1)/4 + \sum_{a \succ_W b} w_D(b \to a). \tag{3.13}$$

Therefore, a minimizer of $\text{KT}(W, D)$ is a minimizer of $\sum_{a \succ_W b} w_D(b \to a)$, which is the total weight of edges in the WMG that are not consistent with W. It follows that these edges correspond to a minimum-weight feedback arc set. $\square$

Example 3.25   Let D denote the preference profile in Figure 3.4. The unique minimum feedback arc set is $\{c \to d\}$. Kemeny outputs two linear orders: $[d \succ a \succ b \succ c]$ and $[d \succ b \succ a \succ c]$, which are the linearizations of the remaining WMG after removing $c \to d$.

In light of Theorem 3.24 and the NP-hardness of minimum FEEDBACK ARC SET, it is not hard to see that the MLE under Mallows is NP-hard to compute [Bartholdi et al., 1989]. The following integer linear program (ILP) by Conitzer et al. [2006] computes the MLE under Mallows, which is the outcome of the Kemeny rule. For each pair of different alternatives $a, b \in \mathcal{A}$, there is a binary variable x_{ab} that corresponds to the pairwise comparison between a and b:

$$\max \sum_{a,b \in \mathcal{A}} w_D(a \to b)x_{ab}$$
$$s.t. \quad \text{for all } a \neq b : x_{ab} + x_{ba} = 1$$
$$\text{for all pairwise different } a, b, c : x_{ab} + x_{bc} + x_{ca} \leq 2.$$

The goal is to compute a linear order to maximize negative Kendall–Tau distance, which amounts to maximizing $\sum_{a,b \in \mathcal{A}} w_D(a \to b)x_{ab}$ as in the ILP. Two types of constraints are included to guarantee that x_{ab}'s correctly encode a linear order. Note that the second type of constraints only prevent "cycles" of length 3, which suffices because it can be proved that if the conditions hold for all triples of alternatives, then x_{ab}'s must encode a linear order. MLE under Mallows is therefore computed by translating the optimal solutions of the ILP to linear orders.

MLE under Condorcet's model is easy to compute. Notice that (3.13) also holds when W is a binary relation. Therefore, the binary relations with maximum likelihood are exactly the binary relations that are compatible with the positive edges in the WMG.

Example 3.26   Let D denote the preference profile in Figure 3.4. The MLE under Condorcet's model outputs two binary relations shown in Figure 3.5. Both binary relations are compatible with Figure 3.4. The difference is in the pairwise comparison between $\{a, b\}$: $a \succ b$ in Figure 3.2a, and $b \succ a$ in Figure 3.2b .

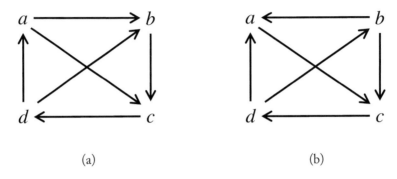

(a)   (b)

Figure 3.5: MLE under Condorcet's model for D in Figure 3.4.

3.4   BIBLIOGRAPHICAL NOTES

Section 3.1.   Condition 3.3 was proposed by Ford, Jr. [1957]. Section 3.1.1 is based on Hunter [2004]. Section 3.1.2 is based on Maystre and Grossglauser [2015]. The LSR algorithm was originally designed for choice data and can be applied to linear orders as discussed in Section 2.2.4. Algorithm 3.6 is based on the codes by Maystre and Grossglauser [2015] and is slightly different from the LSR algorithm described in their paper. An accelerated LSR algorithm was proposed in a subsequent work [Agarwal et al., 2018]. Section 3.1.3 is based on Azari Soufiani et al. [2013a].

Section 3.2.   Sections 3.2.1–3.2.3 are based on Azari Soufiani et al. [2012]. Section 3.2.4 is based on Azari Soufiani et al. [2014a]. GMM was proposed by Hansen [1982].

Section 3.3.   The NP-hardness of Kemeny was proved by Bartholdi et al. [1989]. Hemaspaandra et al. [2005] proved that the problem is $P_{||}^{NP}$-complete. The ILP formulation was proposed by Conitzer et al. [2006]. Approximation algorithms [Ailon et al., 2008] and fixed-parameter algorithms [Betzler et al., 2009] have been designed. Kenyon-Mathieu and Schudy [2007] proposed a polynomial-time approximation scheme (PTAS) for computing Kemeny. Ali and Meila [2012] surveyed many practical algorithms for computing Kemeny.

CHAPTER 4

The Rank-Breaking Framework

The two GMM algorithms (Algorithms 3.7 and 3.10) in the last chapter share the same pattern shown in Figure 4.1.

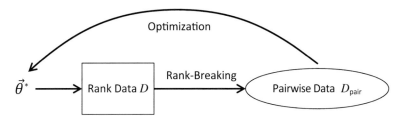

Figure 4.1: Overview of the rank-breaking framework.

As shown in Figure 4.1, the two algorithms first build a dataset of pairwise comparisons D_{pair} from the dataset D of linear orders, by applying a rank-breaking method. Then, a GMM algorithm is applied to D_{pair} for parameter estimation. The advantage of this approach is that parameter estimation from D_{pair} is often computationally easier than parameter estimation from D directly.

In this chapter, we generalize this idea to the *rank-breaking* framework, which can be used to design new algorithms for learning from rank data. The rank-breaking framework is quite general: the algorithm designer can choose any rank-breaking method and any optimization method for pairwise data D_{pair}. As we will see in Section 4.1.4, the framework provides a natural way to explore the tradeoff between computational efficiency and statistical efficiency. We will focus on rank-breaking methods that are characterized by undirected graphs over positions $\{1, \ldots, m\}$ in a linear order, defined as follows.

Definition 4.1 Unweighted Rank-Breaking [Azari Soufiani et al., 2013a]. An unweighted *(rank-)breaking* G is a non-empty undirected graph whose nodes are $\{1, \ldots, m\}$. For any $R \in \mathcal{L}(\mathcal{A})$, we define $G(R)$ as follows:

$$G(R) = \{R[i] \succ R[k] : i < k \text{ and } \{i, k\} \text{ is an edge in } G\}.$$

For any preference profile D, let $G(D) = \bigcup_{R \in D} G(R)$, where $G(D)$ is a multi-set.

In other words, $G(R)$ contains all pairwise comparisons $a \succ b$ that satisfy two conditions: (1) a is preferred to b in R and (2) there is an edge in G between a's rank in R and b's rank in R. We note that nodes in G are the m positions in R.

In this section, we will focus on unweighted breakings. Weighted breakings will be explored in Section 4.2. Let us see a few examples of natural unweighted breakings.

Example 4.2 Figure 4.2 shows the following unweighted breakings.

- **Full breaking:** G_{Full} is the complete graph.

- **Top-k breaking for $k \leq m$:** $G_{\text{Top}}^{k} = \{\{i, j\} : i \leq k, j \neq i\}$.

- **Bottom-k breaking for $k \geq 2$:** $G_{\text{Btm}}^{k} = \{\{i, j\} : i, j \geq m + 1 - k, j \neq i\}$.

- **Adjacent breaking:** $G_{\text{Adj}} = \{\{1, 2\}, \{2, 3\}, \ldots, \{m - 1, m\}\}$.

- **Position-k breaking for $k \geq 2$:** $G_{\text{Pos}}^{k} = \{\{k, i\} : i > k\}$.

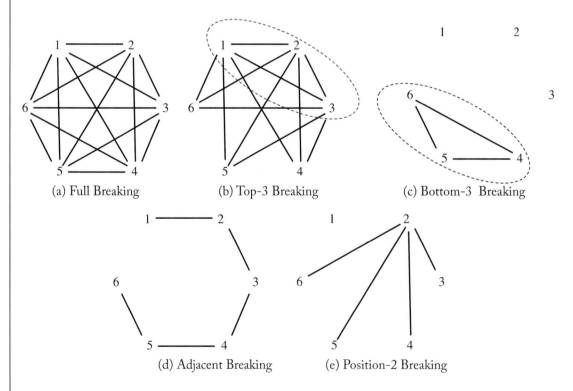

(a) Full Breaking (b) Top-3 Breaking (c) Bottom-3 Breaking

(d) Adjacent Breaking (e) Position-2 Breaking

Figure 4.2: Some unweighted breakings for $m = 6$.

In other words, the full breaking extracts all pairwise comparisons in a linear order. The top-k breaking extracts all pairwise comparisons within top-k alternatives in a linear order, and their comparisons with the remaining alternatives. The bottom-k breaking extracts all pairwise comparisons within bottom-k alternatives. The adjacent breaking extracts pairwise comparisons between $m - 1$ pairs of adjacent alternatives. The position-k breaking extracts pairwise comparisons between the alternative ranked at k-th position in a linear order and all alternatives ranked below it.

Example 4.3   Let $D = \{2@[a \succ b \succ c], 1@[c \succ b \succ a]\}$. We have:

- $G_{\text{Full}}(D) = G_{\text{Top}}^2(D) = G_{\text{Top}}^3(D) = G_{\text{Btm}}^3(D)$

  $= \{2@[a \succ b], 2@[a \succ c], 2@[b \succ c], 1@[c \succ b], 1@[c \succ a], 1@[b \succ a]\}$,

- $G_{\text{Top}}^1(D) = G_{\text{Pos}}^1(D) = \{2@[a \succ b], 2@[a \succ c], 1@[c \succ b], 1@[c \succ a]\}$, and

- $G_{\text{Pos}}^2(D) = G_{\text{Btm}}^2(D) = \{2@[b \succ c], 1@[b \succ a]\}$.

Chapter Overview.   Section 4.1 focuses on the combination of unweighted breaking + GMM algorithms. We will see conditions for the proposed algorithms to output a unique solution and conditions for them to be consistent, for Plackett–Luce (Section 4.1.1) and for general RUMs (Section 4.1.5). Section 4.2 introduces the rank-breaking-then-composite-marginal-likelihood (RBCML) framework, which is the combination of weighted breaking + composite marginal likelihood methods. We will see conditions for RBCML algorithms to satisfy fundamental properties such as consistency and asymptotic normality. Concavity of the objective function of RBCML and consistency of RBCML will be analyzed in detail for Plackett–Luce (Section 4.2.5) and for general RUMs with location families (Section 4.2.6).

4.1   RANK-BREAKING FOR RANDOM UTILITY MODELS

Let us start with designing a GMM algorithm based on D_{pair} to illustrate the idea. To this end, we need to specify moment conditions and make sure that global identification (Condition 3.10) is satisfied.

Given an RUM and a breaking G, for any pair of alternatives a and b, and any parameter $\vec{\theta}$, let $\text{Pr}_{\vec{\theta}, G}(a \succ b)$ denote the probability for $a \succ b$ to appear in $G(R)$, where R is generated from $\pi_{\vec{\theta}}$. More precisely,

$$\text{Pr}_{\vec{\theta}, G}(a \succ b) = \pi_{\vec{\theta}}(\{R : [a \succ b] \in G(R)\}).$$

It is possible that $\text{Pr}_{\vec{\theta}, G}(a \succ b) + \text{Pr}_{\vec{\theta}, G}(b \succ a) < 1$, because G may not be the complete graph, which means that it is possible that neither $a \succ b$ nor $b \succ a$ is in the outcome of the breaking.

Now, define

$$X_G^{a \succ b}(R) = \begin{cases} 1 & [a \succ b] \in G(R) \\ 0 & \text{otherwise.} \end{cases}$$

We will use the following moment condition for each pair of alternatives (a, b) in the rest of this section:

$$g_G^{ab}(R, \vec{\theta}) = X_G^{a \succ b}(R) \times \Pr\nolimits_{\vec{\theta}, G}(b \succ a) - X_G^{b \succ a}(R) \times \Pr\nolimits_{\vec{\theta}, G}(a \succ b). \tag{4.1}$$

The challenge of applying the GMM algorithm with the moment conditions in (4.1) is that the objective function $\sum_{a,b}(g_G^{ab}(D, \vec{\theta}))^2$ can be hard to optimize. This is addressed by using *consistent breakings*, defined as follows.

Definition 4.4 Consistent Breakings.   A breaking G is *consistent* for a RUM, if for any parameter $\vec{\theta}$ and any pair of alternatives a and b, we have

$$\frac{\Pr_{\vec{\theta}, G}(a \succ b)}{\Pr_{\vec{\theta}, G}(b \succ a)} = \frac{\Pr_{\vec{\theta}, G_{\text{Full}}}(a \succ b)}{\Pr_{\vec{\theta}, G_{\text{Full}}}(b \succ a)} = \frac{\pi_{\vec{\theta}}(a \succ b)}{\pi_{\vec{\theta}}(b \succ a)}. \tag{4.2}$$

It follows from the definition that the full breaking is consistent. As we will see soon, in many cases the consistency of G implies the consistency of the GMM algorithm with the moment conditions (4.1) and also simplifies the optimization problem.

4.1.1   BREAKING + GMM FOR PLACKETT–LUCE

In this section, we adopt the $\vec{\gamma}$ parameterization of Plackett–Luce (Definition 2.14). Recall from Equation (3.4) that under Plackett–Luce,

$$\frac{\pi_{\vec{\gamma}}(a \succ b)}{\pi_{\vec{\gamma}}(b \succ a)} = \frac{\gamma_a}{\gamma_b}.$$

Therefore, for any consistent breaking G, (4.1) is equivalent to the following moment condition:

$$g_G^{ab}(R, \vec{\gamma}) = X_G^{a \succ b}(R) \times \gamma_b - X_G^{b \succ a}(R) \times \gamma_a. \tag{4.3}$$

This gives us a GMM algorithm that computes $\vec{\gamma}$ to minimize $\sum_{a \neq b} \left(g_G^{ab}(D, \vec{\gamma}) \right)^2$. The next theorem says that its optimization problem is easier than it appears—$\vec{\gamma}$ can be computed by solving $\vec{\gamma} \times P(G(D)) = \vec{0}$, where $P(\cdot)$ is the matrix computed by counting pairwise comparisons as defined in (3.5).

Theorem 4.5 [Azari Soufiani et al., 2013a].   *For any breaking G and any preference profile D, there exists $\vec{\gamma} \in [0, 1]^m$ with $\vec{\gamma} \cdot \vec{1} = 1$, such that $\vec{\gamma} \times P(G(D)) = \vec{0}$.*

Theorem 4.5 leads to the GMM_{PL}^G algorithm (Algorithm 4.11).

Algorithm 4.11 GMM_{PL}^G (rank-breaking algorithm for Plackett–Luce) [Azari Soufiani et al., 2013a].

Input: A preference profile D and a breaking G.
Output: Estimated parameters of Plackett–Luce.

  1: Compute positive solutions $\vec{\Gamma}$ of $\vec{\gamma} \times P(G(D)) = \vec{0}$ according to Definition 3.7.
  2: **return** $\vec{\Gamma}$

GMM_{PL} (Algorithm 3.7) for Plackett–Luce in Section 3.1 is a special case of GMM_{PL}^G, where $G = G_{\text{Full}}$. The following theorem says that the consistency of G is equivalent to the consistency of GMM_{PL}^G, which is proved by verifying the sufficient conditions for GMM to be consistent [Hansen, 1982].

Theorem 4.6   Consistency of Algorithm 4.11 [Azari Soufiani et al., 2013a].   *GMM_{PL}^G is consistent for the Plackett–Luce model if and only if G is consistent.*

4.1.2   UNIQUENESS OF OUTCOME OF ALGORITHM 4.11

It is possible that $\text{GMM}_{\text{PL}}^G(D)$ is empty or multi-valued for some data D. Let us take a closer look at $P(G(D))$ (Definition 3.7). For each $i \leq m$, the (i, i) entry of $P(G(D))$ is negative, all other entries are non-negative, and the absolute value of all entries are bounded above by m. We further recall that $P(G(D)) \times (\vec{1})^\top = 0$. This means that $I + \frac{P(G(D))}{m}$ is a right stochastic matrix, and

$$\vec{\gamma} \times \left(I + \frac{P(G(D))}{m} \right) = \vec{\gamma}.$$

Therefore, any non-zero non-negative solution $\vec{\gamma}$ to the equation above is a stationary distribution of the Markov chain whose transition matrix is $I + \frac{P(G(D))}{m}$. Recall that $\vec{\gamma}$ is a parameter of Plackett–Luce, which means that it must be strictly positive. Therefore, the uniqueness of $\text{GMM}_{\text{PL}}^G(D)$ is equivalent to the uniqueness of strictly positive stationary distributions of the Markov chain with transition matrix $I + \frac{P(G(D))}{m}$.

The next theorem reveals relationships among three conditions, which can be used to characterize D for which $|\text{GMM}_{\text{PL}}^G(D)| = 1$. Recall that the transition matrix M of a Markov chain is *irreducible*, if any state can be reached from any other state. That is, M only has one communicating class.

Theorem 4.7 [Azari Soufiani et al., 2013a].   *The following conditions are equivalent for any data D and any breaking G.*

  1. $G(D)$ is well-connected.

   2. $I + \frac{P(G(D))}{m}$ is irreducible.

   3. $|GMM_{PL}^{G}(D)| = 1$.

Proof. It is not hard to see that 1 and 2 are equivalent.

   $2 \Rightarrow 3$: By Levin et al. [2008, Proposition 1.14], if $I + P(G(D))/m$ is irreducible, then $I + P(G(D))/m$ has a unique strictly positive stationary distribution, which means that $|\text{GMM}_{\text{PL}}^{G}(D)| = 1$.

   $3 \Rightarrow 2$: suppose for the sake of contradiction that $I + P(G(D))/m$ is not irreducible. There are two cases.

   Case 1: There exists an inessential state. Then, for any stationary distribution, the inessential state must have 0 probability [Levin et al., 2008, Proposition 1.25]. This means that $\text{GMM}_{\text{PL}}^{G}(D) = \emptyset$.

   Case 2: There is no inessential state. In this case all essential communicating classes do not communicate. Therefore, any convex combination of their corresponding stationary distributions is a stationary distribution of the Markov chain. This means that $|\text{GMM}_{\text{PL}}^{G}(D)| = \infty$. $\qquad\square$

   When G is the complete graph, $G_{\text{Full}}(D)$ is well-connected if and only if D is well-connected (Condition 3.3). Because GMM_{PL} (Algorithm 3.7) is $\text{GMM}_{\text{PL}}^{G_{\text{Full}}}$, we have the following corollary of Theorem 4.7.

Corollary 4.8 *For any preference profile D, $|GMM_{PL}(D)| = 1$ if and only if D is well-connected.*

4.1.3   CHARACTERIZATION OF CONSISTENT BREAKINGS FOR PLACKETT–LUCE

Theorem 4.6 shows that consistent breakings are desirable for $\text{GMM}_{\text{PL}}^{G}$. The following theorem characterizes all consistent breakings for Plackett–Luce.

Theorem 4.9   Consistent Breakings for Plackett–Luce [Azari Soufiani et al., 2013a, 2014a]. *A breaking G is consistent for the Plackett–Luce model if and only if G is the union of position-k breakings.*

Proof idea: The idea behind the $\Leftarrow$ direction is the following. First, it can be shown that any position-k breaking is consistent, by directly calculating the probabilities of pairwise comparisons and verifying Equation (4.2) in Definition 4.4. The $\Leftarrow$ direction follows immediately from the following lemma.

Lemma 4.10   Consistent Breaking ± Consistent Breaking.   *Let G_1, G_2 be a pair of consistent breakings.*

   1. If $G_1 \cap G_2 = \emptyset$, then $G_1 \cup G_2$ is also consistent.

2. *If $G_1 \subsetneq G_2$ and $(G_2 \setminus G_1) \neq \emptyset$, then $(G_2 \setminus G_1)$ is also consistent.*

The $\Rightarrow$ direction is proved by induction on m and the following lemma.

Lemma 4.11 Consistent Breaking ± Inconsistent Breaking. *Let G_1 be consistent and let G_2 be inconsistent, then*

1. *if $G_1 \cap G_2 = \emptyset$, then $G_1 \cup G_2$ is inconsistent;*

2. *if $G_1 \subsetneq G_2$, then $(G_2 \setminus G_1)$ is inconsistent; and*

3. *if $G_2 \subsetneq G_1$, then $(G_1 \setminus G_2)$ is inconsistent.*

It is not hard to verify that the $\Rightarrow$ directly holds for $m = 3$. For any $1 \leq k_1 < k_2 \leq m$, let $G_{[k_1,k_2]}$ denote the subgraph of G for positions in $[k_1, k_2]$. We subtract $k_1 - 1$ from each node in $G_{[k_1,k_2]}$, so that they become $[1, \ldots, k_2 - k_1 + 1]$.

Suppose the theorem holds for l. When $m = l + 1$, it can be shown by induction hypothesis that $G_{[2,m]}$ must be the union of a set of position-k breakings. The same argument holds for $G_{[1,m-1]}$. Then, there are two cases.

- **Case 1:** $G_{[1,m-1]}$ contains the position-1 breaking, which means that for all $i \leq m - 1$, $\{1, i\} \in G$. We claim that $\{1, m\} \in G$. First, we note that $\{1, m\} \cup G$ is consistent because it is the union of position-1 breaking and other position-k breakings, which implies consistency by the $\Leftarrow$ direction. Second, it can be proved that $\{1, m\}$ is inconsistent.

  Then, by Lemma 4.11, $(G \setminus \{1, m\})$ is inconsistent. It follows that $\{1, m\} \in G$ because G is consistent.

- **Case 2:** $G_{[1,m-1]}$ does not contain the position-1 breaking, which means that for all $i \leq m - 1$, $\{1, i\} \notin G$. In this case we can prove $\{1, m\} \notin G$ following a similar argument.

In both cases the $\Rightarrow$ direction holds for $m = l + 1$, which proves the theorem.   □

Example 4.12 Union of Position-k Breakings.   Among the breakings in Example 4.2:

- the full breaking G_{Full} is the union of position-1 through position-$(m - 1)$ breakings;

- top-k breaking G_{Top}^k is the union of position-1 through position-k breakings;

- bottom-k breaking G_{Btm}^k is the union of position-k through position-$(m - 1)$ breakings; and

- for any $m \geq 3$, the adjacent breaking G_{Adj} cannot be represented as the union of any combination of position-k breakings.

Therefore, we have the following corollary of Theorem 4.9.

Corollary 4.13   *The full breaking G_{Full}, top-k breaking G^k_{Top}, position-k breaking G^k_{Pos}, and bottom-k breaking G^k_{Btm} are consistent. For any $m \geq 3$, the adjacent breaking G_{Adj} is inconsistent.*

The corollary suggests that for parameter estimation under Plackett–Luce, we should use consistent breakings such as the four types of consistent breakings mentioned in Corollary 4.13. The adjacent breaking is natural but inconsistent, which means that it does not provide a good estimation for Plackett–Luce.

4.1.4   COMPUTATIONAL AND STATISTICAL EFFICIENCY OF ALGORITHMS FOR PLACKETT–LUCE

Table 4.1 summarizes the computational complexity of five parameter estimation algorithms for Plackett–Luce: the MM algorithm and the LSR algorithm in Chapter 3, and GMM^G_{PL} with the full breaking, top-k breaking, and adjacent breaking, respectively.

Table 4.1: Computational complexity of parameter estimation algorithms for Plackett–Luce [Azari Soufiani et al., 2013a]

Algorithm	Breaking	Optimization	Overall
MM (Algorithm 3.5)	N/A	$\Theta(m^3 n)$ per iteration	$\Theta(m^3 n)$ per iteration
LSR (Algorithm 3.6)	N/A	$\Theta(m^2 n) + O(m^{2.376})$ per iteration	$\Theta(m^2 n) + O(m^{2.376})$ per iteration
$\text{GMM}^{G_{\text{Full}}}_{\text{PL}}$ (Algorithm 4.11, full breaking)	$\Theta(m^2 n)$	$O(m^{2.376})$	$O(m^2 n + m^{2.376})$
$\text{GMM}^{G^k_{\text{Top}}}_{\text{PL}}$ (Algorithm 4.11, top-k breaking)	$\Theta(mkn)$	$O(m^{2.376})$	$O((m+k)kn + m^{2.376})$
$\text{GMM}^{G_{\text{Adj}}}_{\text{PL}}$ (Algorithm 4.11, adjacent breaking)	$\Theta(mn)$	$O(m^{2.376})$	$O(mn + m^{2.376})$

The $O(m^{2.376})$ runtime for optimization in GMM^G_{PL} and LSR comes from solving a set of m linear equations.

Computational Efficiency.   Table 4.1 suggests the following ranking w.r.t. asymptotic runtime (where $\succ$ means "better than"):

$$\text{Runtime:}\quad \text{GMM}^{G_{\text{Adj}}}_{\text{PL}} \succeq \text{GMM}^{G^k_{\text{Top}}}_{\text{PL}} \succ \text{GMM}^{G_{\text{Full}}}_{\text{PL}} \approx \text{LSR} \succ \text{MM}.$$

The first $\succeq$ is not strict because when $k = 1$, the asymptotic complexity of $\text{GMM}_{\text{PL}}^{G_{\text{Adj}}}$ and the asymptotic complexity of $\text{GMM}_{\text{PL}}^{G_{\text{Top}}^k}$ are the same. This relationship is justified by experiments on synthetic data. The comparison between $\text{GMM}_{\text{PL}}^{G_{\text{Full}}}$ and MM can be found in Azari Soufiani et al. [2013a]. The comparisons of $\text{GMM}_{\text{PL}}^{G_{\text{Full}}}$, LSR (one-round LSR, namely $T = 2$ in Algorithm 3.6), and 2-LSR (two-round LSR, namely $T = 3$ in Algorithm 3.6) are shown in Figure 4.3a.

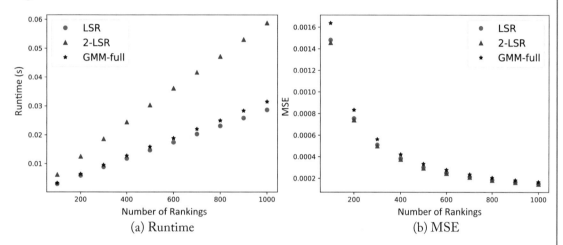

(a) Runtime   (b) MSE

Figure 4.3: Runtime and MSE of LSR, 2-LSR, and $\text{GMM}_{\text{PL}}^{G_{\text{Full}}}$ for Plackett–Luce. In the experiments $m = 10$, n ranges from 100–1000, $\vec{\gamma}$ is generated uniformly at random and normalized, and the results are calculated by 1000 trials.

Statistical Efficiency. Because $\text{GMM}_{\text{PL}}^{G_{\text{Adj}}}$ is inconsistent while the other algorithms mentioned above are consistent, as data size grows, $\text{GMM}_{\text{PL}}^{G_{\text{Adj}}}$ has poor accuracy compared to other methods. Experiments on synthetic data suggest the following order w.r.t. statistical efficiency measured by the mean squared error (MSE):

$$\text{MSE:}\quad \text{LSR} \succ \text{GMM}_{\text{PL}}^{G_{\text{Full}}} \succ \text{GMM}_{\text{PL}}^{G_{\text{Top}}^k} \succ \text{GMM}_{\text{PL}}^{G_{\text{Adj}}}.$$

The MSE of LSR, 2-LSR, and $\text{GMM}_{\text{PL}}^{G_{\text{Full}}}$ are shown in Figure 4.3b. The one-round LSR has similar runtime and lower MSE compared to $\text{GMM}_{\text{PL}}^{G_{\text{Full}}}$. In fact, the one-round LSR can be seen as a variant of GMM_{PL}, where $P(D)$ is replaced by the outcome of a weighted breaking that will be introduced in Section 4.2.1. See Example 4.20 for more details.

   If MM were to run until full convergence, which requires significantly longer time than other algorithms mentioned above, then its MSE would be as good as the MSE of LSR, because both algorithms converge to the MLE.

Runtime vs. MSE Tradeoff among Top-k Breakings.   For GMM algorithms with top-k breakings, the larger k is, the more information is utilized from data at a higher computational cost. Therefore, it is natural to conjecture that $\mathrm{GMM}_{\mathrm{PL}}^{G_{\mathrm{Top}}^k}$ with larger k has lower MSE compared to $\mathrm{GMM}_{\mathrm{PL}}^{G_{\mathrm{Top}}^k}$ with smaller k. In other words, we expect to see a runtime-MSE tradeoff among $\mathrm{GMM}_{\mathrm{PL}}^{G_{\mathrm{Top}}^k}$ for different k's. This tradeoff is verified by the experiment results in Figure 4.4.

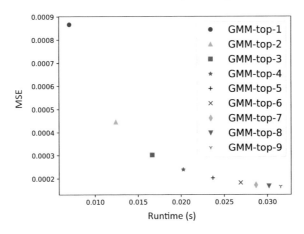

Figure 4.4: Runtime vs. MSE tradeoff among top-k breakings under Plackett–Luce. In the experiments $m = 10$, $n = 1000$, $\vec{\gamma}$ is generated uniformly at random and normalized, and the results are calculated by 1000 trials.

4.1.5 RANK-BREAKING FOR GENERAL RANDOM UTILITY MODELS WITH LOCATION FAMILIES

In this section, we extend $\mathrm{GMM}_{\mathrm{RUM}}$ (Algorithm 3.10) to a GMM algorithm with an arbitrary breaking G for general RUMs with location families. Recall that in an RUM with location families, a parameter is denoted by $\vec{\theta} = (\theta_1, \ldots, \theta_m)$, where θ_i represents the mean of the utility distribution for alternative a_i. Also recall that θ_m is always set to be 0 to guarantee that the model is identifiable (Property 2.23).

Recall that in $\mathrm{GMM}_{\mathrm{RUM}}$, there is a moment condition g^{ab} for each pair of different alternatives a and b:

$$g^{ab}(R, \vec{\theta}) = X^{a \succ b}(R) \times f^{ba}(\theta_b - \theta_a) - X^{b \succ a}(R) \times f^{ab}(\theta_a - \theta_b),$$

where $f^{ba}(\theta_b - \theta_a)$ is the probability for $b \succ a$ under full breaking as defined in (3.11). For any consistent breaking G, (4.1) becomes:

$$g_G^{ab}(R, \vec{\theta}) = X_G^{a \succ b}(R) \times f^{ba}(\theta_b - \theta_a) - X_G^{b \succ a}(R) \times f^{ab}(\theta_a - \theta_b). \tag{4.4}$$

This leads to the $\text{GMM}^G_{\text{RUM}}$ algorithm (Algorithm 4.12).

Algorithm 4.12 $\text{GMM}^G_{\text{RUM}}$ (rank-breaking algorithm for RUMs with location families).

Input: A preference profile D, an RUM with location families, and a breaking G.
Output: Estimated parameter of the RUM.

1: For all pairs of alternatives $a \neq b$, compute $X^{a \succ b}_G(D)$.
2: Compute $\vec{\theta}^* \in \arg\min_{\vec{\theta}} \sum_{a \neq b} g^{ab}_G(D, \vec{\theta})^2$.
3: **return** $\vec{\theta}^*$.

The following theorem gives a sufficient condition for $\text{GMM}^G_{\text{RUM}}$ to be consistent.

Theorem 4.14   Consistency of $\text{GMM}^G_{\text{RUM}}$ [Azari Soufiani et al., 2014a].   *For any RUM with location families and any consistent breaking G, if the PDF of every utility distribution π^0_i is continuous, then $\text{GMM}^G_{\text{RUM}}$ (Algorithm 4.12) is consistent.*

Recall that π^0_i is the zero-mean distribution for a_i. The next theorem shows that, unfortunately, the full breaking is the only consistent breaking for many RUMs with location families where the PDFs of the utility distributions are symmetric around their means.

Theorem 4.15   Inconsistent Breakings [Azari Soufiani et al., 2014a].   *Given an RUM with location families, if the PDF of each utility distribution π^0_i has support on $(-\infty, \infty)$ and is symmetric around its mean, then the only consistent breaking is the full breaking.*

We note that Theorem 4.15 does not apply to Plackett–Luce because the PDF of Gumbel is not symmetric around its mean. The proof is based on the following theorem, which applies to all RUMs with location families, where π^0_i's are not required to be symmetric.

Theorem 4.16   [Azari Soufiani et al., 2014a].   *For any RUM with location families where each utility distribution has support $(-\infty, \infty)$, if the full breaking is the only consistent breaking for all sub-models with three alternatives,[1] then the full breaking is the only consistent breaking for any m.*

By Theorem 4.16, in order to check whether the full breaking is the only consistent breaking, it suffices to check whether there are any consistent breakings beyond the full breaking for $m = 3$. The answer is no for RUMs with symmetric utility distributions, which proves Theorem 4.15.

We note that Theorem 4.16 also applies to Plackett–Luce. Because the position-2 breaking for $m = 3$ is consistent under Plackett–Luce, Theorem 4.16 does not imply that full breaking is the only consistent breaking under Plackett–Luce. This agrees with the characterization of consistent breakings under Plackett-Luce (as unions of position-k breakings) in Theorem 4.9.

---

[1]A sub-model of $\text{RUM}(\mathcal{M}_1, \ldots, \mathcal{M}_m)$ is $\text{RUM}(\mathcal{M}_{i_1}, \ldots, \mathcal{M}_{i_i})$ for some $\{i_1, \ldots, i_k\} \subseteq \{1, \ldots, m\}$.

4.2 RANK-BREAKING + COMPOSITE MARGINAL LIKELIHOOD (RBCML)

Recall from Figure 4.1 that the rank-breaking framework has two components: a rank-breaking method and an optimization method. Section 4.1 focused on the combination of unweighted breaking and GMM. A different combination will be explored in this section: *weighted breaking*, which is a natural extension of unweighted breaking, and *composite marginal likelihood methods (CML)*, which is a natural extension of MLE.

4.2.1 WEIGHTED BREAKINGS

Let us start with the definition of weighted breakings.

Definition 4.17 Weighted Breakings [Khetan and Oh, 2016]. A *weighted (rank-)breaking* $\mathcal{G} = \{v_{ik} : i, k \leq m \text{ and } v_{ik} = v_{ki}\}$ is a weighted undirected graph over positions $\{1, \ldots, m\}$, such that for any $v_{ik} > 0$, there is an edge between i and k whose weight is v_{ik}; if there is no edge between i and k, then $v_{ik} = 0$.

A weighted breaking $\mathcal{G}$ maps a linear order $R \in \mathcal{L}(\mathcal{A})$ to weighted pairwise comparisons, defined as follows.

Definition 4.18 For any weighted breaking $\mathcal{G}$, any pair of different alternatives (a_{i_1}, a_{i_2}), and any linear order $R \in \mathcal{L}(\mathcal{A})$,

- suppose a_{i_1} and a_{i_2} are ranked at the i-th position and the k-th position in R, respectively, then let

$$\kappa_{i_1 i_2}(R) = \begin{cases} v_{ik} & \text{if } i < k \\ 0 & \text{if } i > k \end{cases};$$

  and

- let $\mathcal{G}(R)$ denote the set of weighted pairwise comparisons, where the weight of $a_{i_1} \succ a_{i_2}$ is $\kappa_{i_1 i_2}(R)$.

  For any preference profile D of n linear orders, let

$$\kappa_{i_1 i_2}(D) = \frac{1}{n} \sum_{R \in D} \kappa_{i_1 i_2}(R) \quad \text{and} \quad \mathcal{G}(D) = \frac{1}{n} \sum_{R \in D} \mathcal{G}(R).$$

Slightly abusing notation, let $\mathcal{G}(D)$ also denote the weighted directed graph where the nodes are the alternatives and the weight on $a_i \to a_k$ is κ_{ik}.

Note that unlike in $\mathcal{G}$, the nodes in $\mathcal{G}(D)$ are the alternatives. Sometimes we use $\kappa_{i_1 i_2}$ to denote $\kappa_{i_1 i_2}(D)$ when D is clear from the context.

Example 4.19 The Harmonic Breaking [Khetan and Oh, 2016]. In the harmonic breaking $\mathcal{G}_H$, for any $i < k \leq m$, $v_{ik} = \frac{1}{m-i}$. For example, Figure 4.5a shows the harmonic breaking for

$m = 4$. We also have:

$$\mathcal{G}_H(a \succ b \succ c \succ d)$$
$$= \left\{ \frac{1}{3}@[a \succ b], \frac{1}{3}@[a \succ c], \frac{1}{3}@[a \succ d], \frac{1}{2}@[b \succ c], \frac{1}{2}@[b \succ d], 1@[c \succ d] \right\}.$$

Note that $\mathcal{G}_H = \bigcup_{k=1}^{m-1} \frac{1}{m-k} G_{\text{Pos}}^k$. That is, the harmonic breaking is the weighted union of a set of unweighted position-k breakings.

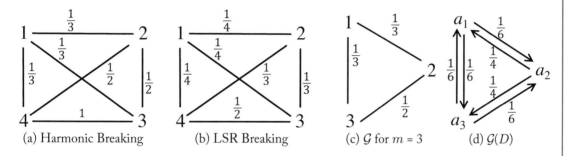

(a) Harmonic Breaking   (b) LSR Breaking   (c) $\mathcal{G}$ for $m = 3$ (d) $\mathcal{G}(D)$

Figure 4.5: Weighted breakings for Examples 4.19, 4.20, and 4.21.

Example 4.20   LSR Breaking [Khetan and Oh, 2016][2]   The one-round LSR algorithm ($T = 2$ in Algorithm 3.6) can be seen as GMM_{PL} (Algorithm 3.7), where $P(D)$ is replaced by the outcome of the LSR breaking: for any $i < k \le m$, $v_{ik} = \frac{1}{m+1-i}$. For example, Figure 4.5b shows the LSR breaking for $m = 4$.

Example 4.21   Let $m = 3, n = 2$ and let the preference profile be $D = \{[a_1 \succ a_2 \succ a_3], [a_3 \succ a_2 \succ a_1]\}$. Let $\mathcal{G} = \{v_{12} = v_{13} = \frac{1}{3}, v_{23} = \frac{1}{2}\}$ as shown in Figure 4.5c. Then, $\kappa_{12} = (\kappa_{12}(a_1 \succ a_2 \succ a_3) + \kappa_{12}(a_3 \succ a_2 \succ a_1))/2 = (\frac{1}{3} + 0)/2 = \frac{1}{6}$. Similarly,

$$\kappa_{12} = \kappa_{13} = \tfrac{1}{3}/n = \tfrac{1}{6}, \quad \kappa_{23} = \tfrac{1}{2}/n = \tfrac{1}{4},$$
$$\kappa_{32} = \kappa_{31} = \tfrac{1}{3}/n = \tfrac{1}{6}, \quad \kappa_{21} = \tfrac{1}{2}/n = \tfrac{1}{4};$$

$\mathcal{G}(D)$ is shown in Figure 4.5d.

4.2.2   COMPOSITE MARGINAL LIKELIHOOD METHODS (CML)

The intuition behind CML methods is: each pairwise data $a \succ b$ can be seen as an event (Section 2.1.1), and the goal is to compute the parameter $\vec{\theta}$ with maximum marginal likelihood. The

---

[2]Khetan and Oh [2016] analyzed this breaking in some of their theorems.

objective function of CML also uses possibly different weights for different pairs of alternatives, encoded in a *CML weight graph*, defined as follows.

Definition 4.22   A *CML weight graph* W is a directed weighted graph over $\mathcal{A}$, where for all pair of different alternatives (a, b), w_{ab} is the weight over the edge $a \to b$. W is *symmetric*, if for any pair of different alternatives (a, b), we have $w_{ab} = w_{ba}$. W is *uniform*, if the weights on all edges are the same. Let W_u denote the uniform CML weight graph.

To simplify notation, sometimes $w_{a_{i_1} a_{i_2}}$ is denoted by $w_{i_1 i_2}$.

Example 4.23   A symmetric W is shown in Figure 4.6. In W, $w_{12} = w_{21} = 1$ and $w_{23} = w_{32} = 2$, where $w_{ik} = w_{a_i a_k}$.

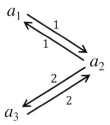

Figure 4.6: Symmetric W.

So far we have defined two types of weights for a pair of alternatives (a_{i_1}, a_{i_2}): $\kappa_{i_1 i_2}$, which is computed by applying breaking $\mathcal{G}$, and $w_{i_1 i_2}$, which is specified in W.

4.2.3   THE RBCML FRAMEWORK

Given a weighted breaking $\mathcal{G}$ and a CML weight graph W, the rank-breaking-then-CML framework for RUMs, denoted by RBCML$(\mathcal{G}, W)$ (Algorithm 4.13), is defined to be the maximizer of the following composite log-marginal likelihood function.

Definition 4.24   Composite (Log-)Marginal Likelihood for RUMs.   Given an RUM and a preference profile D, the composite marginal likelihood is:

$$\mathrm{CL}(\vec{\theta}, D) = \prod_{i_1 \neq i_2} \pi_{\vec{\theta}}(a_{i_1} \succ a_{i_2})^{\kappa_{i_1 i_2} \cdot w_{i_1 i_2}}.$$

The composite log-marginal likelihood is:

$$\mathrm{CLL}(\vec{\theta}, D) = \sum_{i_1 \neq i_2} \kappa_{i_1 i_2} \cdot w_{i_1 i_2} \cdot \ln \pi_{\vec{\theta}}(a_{i_1} \succ a_{i_2}). \tag{4.5}$$

Example 4.25   Consider the Plackett–Luce model with the $\vec{\theta}$ parameterization (Definition 2.16) for $\mathcal{A} = \{a_1, a_2, a_3\}$. Recall that θ_3 is fixed to be 0 for identifiability (Property 2.23). We have:

$$
\begin{aligned}
\text{CLL}(\vec{\theta}, D) &= \sum_{i_1 \neq i_2} \kappa_{i_1 i_2} \cdot w_{i_1 i_2} \cdot \ln \frac{e^{\theta_{i_1}}}{e^{\theta_{i_1}} + e^{\theta_{i_2}}} = \sum_{i_1 \neq i_2} \kappa_{i_1 i_2} \cdot w_{i_1 i_2} \cdot (\theta_{i_1} - \ln(e^{\theta_{i_1}} + e^{\theta_{i_2}})) \\
&= \sum_{i_1 < i_2} \left(\kappa_{i_1 i_2} w_{i_1 i_2} \theta_{i_1} + \kappa_{i_2 i_1} w_{i_2 i_1} \theta_{i_2} - (\kappa_{i_1 i_2} w_{i_1 i_2} \theta_{i_2} + \kappa_{i_2 i_1} w_{i_2 i_1} \theta_{i_2}) \ln(e^{\theta_{i_1}} + e^{\theta_{i_2}}) \right) \\
&= (\kappa_{12} w_{12} + \kappa_{13} w_{13}) \theta_1 + (\kappa_{21} w_{21} + \kappa_{23} w_{23}) \theta_2 - (\kappa_{12} w_{12} + \kappa_{21} w_{21}) \ln(e^{\theta_1} + e^{\theta_2}) \\
&\quad - (\kappa_{13} w_{13} + \kappa_{31} w_{31}) \ln(e^{\theta_1} + 1) - (\kappa_{23} w_{23} + \kappa_{32} w_{32}) \ln(e^{\theta_2} + 1).
\end{aligned}
$$

Algorithm 4.13 RBCML$(\mathcal{G}, \mathcal{W})$.

Input: A preference profile D, an RUM, a weighted breaking $\mathcal{G}$, and a CML weight graph $\mathcal{W}$.
Output: Estimated parameters of the RUM.
  1: For all different pairs of alternatives (a, b), compute κ_{ab} (Definition 4.18).
  2: **return** $\arg\max_{\vec{\theta}} \text{CLL}(\vec{\theta}, D)$ as in (4.5).

Connection to GMM.   CML can be viewed as a special case of GMM in the following way. When the composite log-marginal likelihood is strictly concave, RBCML can be computed by solving the first-order conditions for CLL, which means that the optimization problem can be solved by GMM. As we will see later in this chapter, the composite marginal log-likelihood is strictly concave for Plackett–Luce (Theorem 4.33) and for a natural class of RUMs with location families (Theorem 4.36).

Formally, for any $i \leq m - 1$ (recall that θ_m is always set to be 0), there is a moment condition $g_i(R, \vec{\theta})$ defined as follows:

$$
g_i(R, \vec{\theta}) = \frac{\partial \text{CLL}(\vec{\theta}, R)}{\partial \theta_i} = \sum_{i_1 \neq i_2} \frac{\kappa_{i_1 i_2}(R) \cdot w_{i_1 i_2}}{\pi_{\vec{\theta}}(a_{i_1} \succ a_{i_2})} \cdot \frac{\partial \pi_{\vec{\theta}}(a_{i_1} \succ a_{i_2})}{\partial \theta_i}.
$$

4.2.4   CONSISTENCY AND ASYMPTOTIC NORMALITY OF RBCML

Similar to MLE, under natural conditions, it can be proved that CML methods satisfy desirable statistical properties such as consistency and asymptotic normality. Given RBCML$(\mathcal{G}, \mathcal{W})$ and a parameter $\vec{\theta}^*$, let

$$\text{ELL}(\vec{\theta}^*) = \mathbb{E}_{R \sim \pi_{\vec{\theta}^*}(\cdot)}[\text{CLL}(\vec{\theta}^*, R)]$$

denote the *expected composite log-marginal likelihood* of a single data point that is generated from $\pi_{\vec{\theta}^*}$. Let $\bar{H}(\vec{\theta}^*)$ denote the expected Hessian of $\text{CLL}(\vec{\theta}, R)$ evaluated at $\vec{\theta} = \vec{\theta}^*$, where the expectation is taken for R that is generated from $\pi_{\vec{\theta}^*}$.

Theorem 4.26   Consistency and Asymptotic Normality of RBCML [Zhao and Xia, 2018].
Given an RUM with location families and a parameter $\vec{\theta}^*$, for any $n \in \mathbb{N}$, let D_n denote a profile of n linear orders generated from the RUM given $\vec{\theta}^*$ and let $\vec{\theta}^{(n)} = RBCML(\mathcal{G}, \mathcal{W})(D_n)$. As $n \to \infty$, we have

- **consistency:** $\vec{\theta}^{(n)} \xrightarrow{P} \vec{\theta}^*$ *and*

- **asymptotic normality:** $\sqrt{n}(\vec{\theta}^{(n)} - \vec{\theta}^*) \xrightarrow{d} \mathcal{N}\left(0, \bar{H}^{-1}(\vec{\theta}^*)Cov(\vec{\theta}^*)\bar{H}^{-1}(\vec{\theta}^*)\right)$

if and only if $\vec{\theta}^*$ is the unique solution to

$$\nabla ELL(\vec{\theta}) = \vec{0}, \tag{4.6}$$

where $Cov(\vec{\theta}^*)$ is the covariance matrix of $\nabla CLL(\vec{\theta}, R)$ evaluated at $\vec{\theta} = \vec{\theta}^*$.

Let us look at an example of $\bar{H}$ and $Cov(\vec{\theta}^*)$ for $\mathcal{A} = \{a_1, a_2, a_3\}$, where the CML weight graph is $\mathcal{W}_u$, under the Plackett–Luce model with the $\vec{\theta}$ parameterization. Again, let $\theta_3 = 0$.

Example 4.27   Continuing Example 4.25 with the uniform CML weight graph $\mathcal{W}_u$, for any linear order $R \in \mathcal{L}(\{a_1, a_2, a_3\})$, we have

$$\nabla \text{CLL}(\vec{\theta}, R) = \left[\kappa_{12} - \frac{e^{\theta_1}}{e^{\theta_1}+e^{\theta_2}} + \kappa_{13} - \frac{e^{\theta_1}}{e^{\theta_1}+1}, \kappa_{21} - \frac{e^{\theta_2}}{e^{\theta_1}+e^{\theta_2}} + \kappa_{23} - \frac{e^{\theta_2}}{e^{\theta_2}+1} \right].$$

The Hessian matrix $H(R, \vec{\theta})$ of $\text{CLL}(R, \vec{\theta})$ is

$$H(R, \vec{\theta}) = \begin{bmatrix} -\frac{e^{\theta_1}e^{\theta_2}}{(e^{\theta_1}+e^{\theta_2})^2} - \frac{e^{\theta_1}}{(e^{\theta_1}+1)^2} & \frac{e^{\theta_1}e^{\theta_2}}{(e^{\theta_1}+e^{\theta_2})^2} \\ \frac{e^{\theta_1}e^{\theta_2}}{(e^{\theta_1}+e^{\theta_2})^2} & -\frac{e^{\theta_1}e^{\theta_2}}{(e^{\theta_1}+e^{\theta_2})^2} - \frac{e^{\theta_2}}{(e^{\theta_2}+1)^2} \end{bmatrix}.$$

We note that $H(R, \vec{\theta})$ does not depend on R, which means that $\bar{H}(\vec{\theta}^*) = H(R, \vec{\theta}^*)$. This relationship may not hold for other RUMs. Let $\mathcal{L}(\mathcal{A}) = \{V_j : j \leq 6\}$ denote the set of all six linear orders over $\mathcal{A}$, where:

$$V_1 = [a_1 \succ a_2 \succ a_3], V_2 = [a_1 \succ a_3 \succ a_2], V_3 = [a_2 \succ a_1 \succ a_3],$$
$$V_4 = [a_2 \succ a_3 \succ a_1], V_5 = [a_3 \succ a_1 \succ a_2], V_6 = [a_3 \succ a_2 \succ a_1].$$

We have

$$\text{Cov}(\vec{\theta}^*) = \sum_{j=1}^{6} \pi_{\vec{\theta}^*}(V_j) \left[\nabla \text{CLL}(\vec{\theta}^*, V_j) \right]^{\top} \left[\nabla \text{CLL}(\vec{\theta}^*, V_j) \right], \text{ where}$$

$$\nabla \text{CLL}(\vec{\theta}^*, V_1) = \left[1 - \frac{e^{\theta_1^*}}{e^{\theta_1^*} + e^{\theta_2^*}} + 1 - \frac{e^{\theta_1^*}}{e^{\theta_1^*} + 1}, -\frac{e^{\theta_2^*}}{e^{\theta_1^*} + e^{\theta_2^*}} + 1 - \frac{e^{\theta_2^*}}{e^{\theta_2^*} + 1} \right]$$

$$\nabla \text{CLL}(\vec{\theta}^*, V_2) = \left[1 - \frac{e^{\theta_1^*}}{e^{\theta_1^*} + e^{\theta_2^*}} + 1 - \frac{e^{\theta_1^*}}{e^{\theta_1^*} + 1}, -\frac{e^{\theta_2^*}}{e^{\theta_1^*} + e^{\theta_2^*}} - \frac{e^{\theta_2^*}}{e^{\theta_2^*} + 1} \right]$$

$$\nabla \text{CLL}(\vec{\theta}^*, V_3) = \left[-\frac{e^{\theta_1^*}}{e^{\theta_1^*} + e^{\theta_2^*}} + 1 - \frac{e^{\theta_1^*}}{e^{\theta_1^*} + 1}, 1 - \frac{e^{\theta_2^*}}{e^{\theta_1^*} + e^{\theta_2^*}} + 1 - \frac{e^{\theta_2^*}}{e^{\theta_2^*} + 1} \right]$$

$$\nabla \text{CLL}(\vec{\theta}^*, V_4) = \left[-\frac{e^{\theta_1^*}}{e^{\theta_1^*} + e^{\theta_2^*}} - \frac{e^{\theta_1^*}}{e^{\theta_1^*} + 1}, 1 - \frac{e^{\theta_2^*}}{e^{\theta_1^*} + e^{\theta_2^*}} + 1 - \frac{e^{\theta_2^*}}{e^{\theta_2^*} + 1} \right]$$

$$\nabla \text{CLL}(\vec{\theta}^*, V_5) = \left[1 - \frac{e^{\theta_1^*}}{e^{\theta_1^*} + e^{\theta_2^*}} - \frac{e^{\theta_1^*}}{e^{\theta_1^*} + 1}, -\frac{e^{\theta_2^*}}{e^{\theta_1^*} + e^{\theta_2^*}} - \frac{e^{\theta_2^*}}{e^{\theta_2^*} + 1} \right]$$

$$\nabla \text{CLL}(\vec{\theta}^*, V_6) = \left[-\frac{e^{\theta_1^*}}{e^{\theta_1^*} + e^{\theta_2^*}} - \frac{e^{\theta_1^*}}{e^{\theta_1^*} + 1}, 1 - \frac{e^{\theta_2^*}}{e^{\theta_1^*} + e^{\theta_2^*}} - \frac{e^{\theta_2^*}}{e^{\theta_2^*} + 1} \right].$$

The next example shows the closed-form formula for $\nabla \text{ELL}(\vec{\theta})$.

Example 4.28   For any $i \leq m - 1$, the i-th element of $\nabla \text{ELL}(\vec{\theta})$ under Plackett–Luce is

$$\frac{\partial \text{ELL}(\vec{\theta})}{\partial \theta_i} = \sum_{k \neq i} \left(\frac{\overline{\kappa_{ik}} \cdot w_{ik}}{\pi_{\vec{\theta}}(a_i \succ a_k)} \cdot \frac{\partial \pi_{\vec{\theta}}(a_i \succ a_k)}{\partial \theta_i} + \frac{\overline{\kappa_{ki}} \cdot w_{ki}}{\pi_{\vec{\theta}}(a_k \succ a_i)} \cdot \frac{\partial \pi_{\vec{\theta}}(a_k \succ a_i)}{\partial \theta_i} \right),$$

where $\overline{\kappa_{ik}} = \mathbb{E}_{R \sim \pi_{\vec{\theta}}} \kappa_{ik}(R)$.

4.2.5   RBCML FOR PLACKETT–LUCE

This section focuses on the application of RBCML to Plackett–Luce to answer the following questions.

1. Is CLL for Plackett–Luce strictly concave? (Answered by Theorem 4.33.)

2. Is RBCML for Plackett–Luce consistent? (Answered by Theorem 4.34.)

In this section, we will adopt the $\vec{\theta}$ parameterization of Plackett–Luce (Definition 2.16). By plugging in the likelihood function for Plackett–Luce to (4.5), the composite log-marginal likelihood becomes:

$$\mathrm{CLL}_{\mathrm{PL}}(\vec{\theta}, D) = \sum_{i_1 < i_2} \left(\kappa_{i_1 i_2} \cdot w_{i_1 i_2} \theta_{i_1} + \kappa_{i_2 i_1} \cdot w_{i_2 i_1} \theta_{i_2} \right.$$
$$\left. - \left(\kappa_{i_1 i_2} \cdot w_{i_1 i_2} + \kappa_{i_2 i_1} \cdot w_{i_2 i_1} \right) \ln \left(e^{\theta_{i_1}} + e^{\theta_{i_2}} \right) \right). \qquad (4.7)$$

The first-order conditions are:

$$\forall i \leq m - 1, \frac{\partial \mathrm{CLL}_{\mathrm{PL}}(\vec{\theta}, D)}{\partial \theta_i} = \sum_{k \neq i} \left(\kappa_{ik} \cdot w_{ik} - (\kappa_{ik} \cdot w_{ik} + \kappa_{ki} \cdot w_{ki}) \frac{e^{\theta_i}}{e^{\theta_i} + e^{\theta_k}} \right) = 0.$$

Example 4.29   Continuing Example 4.21 and Example 4.23, we have:

$$\mathrm{CLL}_{\mathrm{PL}}\left(\vec{\theta}, D\right) = \frac{1}{6}\theta_1 + \frac{1}{4}\theta_2 - \left(\frac{1}{6} + \frac{1}{4}\right) \ln\left(e^{\theta_1} + e^{\theta_2}\right) + \frac{1}{2}\theta_2 - \left(\frac{1}{2} + \frac{1}{3}\right) \ln\left(e^{\theta_2} + 1\right).$$

By solving the first-order conditions, we have $e^{\theta_1} = 1$ and $e^{\theta_2} = 1.5$. This means that the outcome of RBCML is $\theta_1 = 0$, $\theta_2 = \ln 1.5$. Recall that θ_3 is set to be 0.

Example 4.30   The K-O Algorithm [Khetan and Oh, 2016]   The algorithm proposed by Khetan and Oh [2016] for linear orders, henceforce K-O algorithm, can be viewed as the RBCML algorithm with harmonic breaking $\mathcal{G}_H$ and $\mathcal{W}_u$.

To analyze strict concavity of $\mathrm{CLL}_{\mathrm{PL}}$ we will use the following definition of weakly and strongly connected graphs.

Definition 4.31   Weakly and Strongly Connected Graphs.   A weighted directed graph is *weakly connected*, if after removing the directions on all edges, the resulting undirected graph is connected. A weighted directed graph is *strongly connected*, if there is a directed path with positive weights between any pair of nodes.

For any pair of weighted graphs G_1 and G_2 over the same nodes, let $G_1 \otimes G_2$ denote the weighted graph, where the weight on each edge $a \to b$ is the multiplication of the weight on $a \to b$ in G_1 and the weight on $a \to b$ in G_2.

Example 4.32   Figure 4.7 shows $G_1 \otimes G_2 = G_3$, where G_1 and G_2 are strongly connected and G_3 is weakly connected. Edges with zero weight are not shown.

Theorem 4.33   Properties of $\mathrm{CLL}_{\mathrm{PL}}(\vec{\theta}, D)$ for Plackett–Luce [Zhao and Xia, 2018].   *For any preference profile D, any weighted breaking $\mathcal{G}$, and any CML weight graph $\mathcal{W}$, we have the following.*

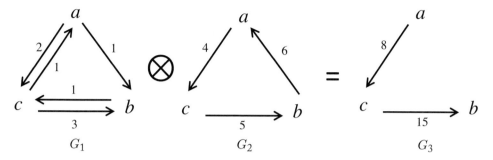

Figure 4.7: $G_1 \otimes G_2 = G_3$.

- **Strict concavity:** $CLL_{PL}(\vec{\theta}, D)$ is strictly concave if and only if $\mathcal{G}(D) \otimes \mathcal{W}$ is weakly connected.

- **Non-Emptiness:** $\arg\max_{\vec{\theta}} CLL_{PL}(\vec{\theta}, D)$ is non-empty if and only if $\mathcal{G}(D) \otimes \mathcal{W}$ is strongly connected.

The following theorem addresses the consistency of RBCML for Plackett–Luce.

Theorem 4.34   Consistency of RBCML for Plackett–Luce [Zhao and Xia, 2018]. *RBCML$(\mathcal{G}, \mathcal{W})$ for Plackett–Luce is consistent if and only if $\mathcal{G}$ is the weighted union of a set of position-k breakings and $\mathcal{W}$ is connected and symmetric.*

Example 4.35   By Theorem 4.34, the K-O algorithm (harmonic breaking + $\mathcal{W}_u$) is consistent. It remains consistent when $\mathcal{W}_u$ is replaced by a graph where for each $1 \leq i \leq m-1$ there is an edge $a_i \rightarrow a_{i+1}$ with weight 1 and an edge $a_{i+1} \rightarrow a_i$ with weight 1.

4.2.6   RBCML FOR RUMS WITH LOCATION FAMILIES

This section answers the same questions for RUMs with location families.

1. Is CLL for RUMs with location families strictly concave? (Answered by Theorem 4.36.)

2. Is RBCML for RUMs with location families consistent? (Answered by Theorems 4.38 and 4.40.)

Theorem 4.36   Properties of CLL$(\vec{\theta}, D)$ for RUMs with Location Families [Zhao and Xia, 2018].   *For any preference profile D, any weighted breaking $\mathcal{G}$, any CML weight graph $\mathcal{W}$, and any RUM with location families, where the CDF of each utility distribution is strictly log-concave, we have the following.*

- **Strict Concavity:** $CLL(\vec{\theta}, D)$ *is strictly concave if and only if* $\mathcal{G}(D) \otimes \mathcal{W}$ *is weakly connected.*

- **Non-Emptiness:** $\arg\max_{\vec{\theta}} CLL(\vec{\theta}, D)$ *is non-empty if and only if* $\mathcal{G}(D) \otimes \mathcal{W}$ *is strongly connected.*

The following lemma provides a sufficient condition for the CDF of a distribution to be strictly log-concave.

Lemma 4.37 [Bagnoli and Bergstrom, 2005] *The CDF of any distribution over* $(-\infty, \infty)$ *with continuously differentiable and strictly log-concave PDF is strictly log-concave.*

For example, the PDF of Gumbel and PDFs of Gaussian distributions are strictly log-concave, which means that Theorem 4.36 can be applied Plackett–Luce and Thurstone's Case V models.

The following theorem provides a necessary and sufficient condition on $\mathcal{G}$ for RBCML$(\mathcal{G}, \mathcal{W}_u)$ to be consistent for a subclass of RUMs with location families.

Theorem 4.38 [Zhao and Xia, 2018]. *Suppose an RUM with location families satisfies: (1) for all* $i \leq m$, π_i^0 *is symmetric; and there exists* i^* *such that (2)* $(\ln \pi_{i^*}(x))'$ *is monotonically decreasing, and (3)* $\lim_{x \to -\infty} (\ln \pi_{i^*}(x))' \to \infty$. *Then, RBCML$(\mathcal{G}, \mathcal{W}_u)$ is consistent if and only if* $\mathcal{G}$ *is the unweighted full breaking.*

Note that Theorem 4.38 holds if there exists an i^* that satisfies condition (2) and (3). It does not require (2) and (3) to hold for all $i \leq m$. Also note that the theorem can be applied to arbitrary combinations of symmetric utility distributions. In particular, we have the following corollary.

Corollary 4.39 *Theorem 4.36 and Theorem 4.38 hold for any RUM with Gaussian distributions of possibility different but fixed variances.*

For any distribution π, we let RUM(π) denote the RUM with location families where each utility distribution has the same shape as π. The following two theorems give stronger characterizations of the consistency of RBCML for RUM(π).

Theorem 4.40 Consistency of RBCML for RUM(π) [Zhao and Xia, 2018]. *Let* π^0 *be any symmetric distribution that satisfies the condition in Theorem 4.38. Then, RBCML$(\mathcal{G}, \mathcal{W})$ is consistent for RUM(π^0) if and only if* $\mathcal{G}$ *is the unweighted full breaking and* $\mathcal{W}$ *is symmetric and (weakly or strongly) connected.*

4.2.7   THE ADAPTIVE RBCML ALGORITHM

As shown in Theorem 4.26, the asymptotic covariance of RBCML depends on $\mathcal{G}$ and $\mathcal{W}$. Ideally we would like to use the optimal $\mathcal{G}$ and $\mathcal{W}$ to minimize the asymptotic MSE, which is the trace of the covariance matrix. However, the optimal $\mathcal{G}$ and $\mathcal{W}$ depend on the ground truth parameter $\vec{\theta}^*$, which is exactly what we want to compute.

The adaptive RBCML algorithm (Algorithm 4.14) is designed to tackle this problem, where $\mathcal{G}$ and $\mathcal{W}$ are iteratively updated by heuristics $\mathcal{G}(\vec{\theta}^{(t)})$ and $\mathcal{W}(\vec{\theta}^{(t)})$, respectively, based on the estimate $\vec{\theta} = \vec{\theta}^{(t)}$ computed in the previous iteration.

Algorithm 4.14 Adaptive RBCML [Zhao and Xia, 2018].

Input: A preference profile D, an RUM with location families, the number of iterations T, $\mathcal{G}(\vec{\theta})$ and $\mathcal{W}(\vec{\theta})$.
Output: Estimated parameter of the RUM.
Initialize $\vec{\theta}^{(1)} = \vec{0}$

1: **for** $t = 1$ **to** T **do**
2:    Compute $\mathcal{G}(\vec{\theta}^{(t)})$ and $\mathcal{W}(\vec{\theta}^{(t)})$.
3:    Choose $\vec{\theta}^{(t+1)} \in \text{RBCML}(\mathcal{G}(\vec{\theta}^{(t)}), \mathcal{W}(\vec{\theta}^{(t)}))$ defined in Algorithm 4.13.
4: **end for**
5: **return** $\vec{\theta}^{(T+1)}$

4.2.8   EXPERIMENTS

In this section, experiment results on synthetic data are shown to compare RBCML algorithms with other algorithms for two RUMs: Plackett–Luce and Thurstone's Case V model (RUMs with Gaussian distributions).

Plackett–Luce with The $\vec{\theta}$ Parameterization.   Figure 4.8 shows the runtime and $n \times$MSE of three algorithms for Plackett–Luce in a synthetic experiment of 50,000 randomly generated preference profiles for $m = 10$. In each profile, each ground truth parameter θ_i is generated uniformly at random from $[0, 5]$.

- **K-O**: is the K-O algorithm, which is the RBCML with harmonic breaking $\mathcal{G}_H$ and $\mathcal{W}_u$ (Example 4.30).

- **RBCML**: is the two-round adaptive RBCML ($T = 2$ in Algorithm 4.14), where the first round uses the K-O algorithm; and the second round uses the harmonic breaking with the following heuristic $\mathcal{W}(\vec{\theta})$. For any pair of alternatives a_{i_1} and a_{i_2}, we let $w_{i_1 i_2} = w_{i_2 i_1} = \frac{1}{|\theta_{i_1} - \theta_{i_2}| + 4}$. The intuition is that closer pairs of alternatives should receive higher weights.

- **1-LSR** is the one-round LSR (Algorithm 3.6).

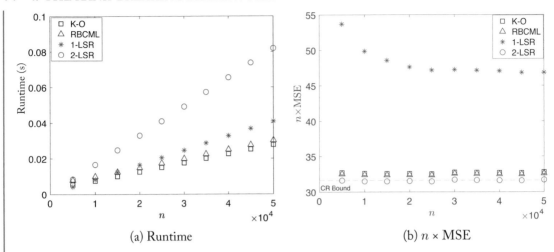

(a) Runtime   (b) $n \times$ MSE

Figure 4.8: The runtime and $n\times$MSE of K-O, adaptive RBCML, 1-LSR, and 2-LSR, for Plackett–Luce.

- **2-LSR** is the two-round LSR (Algorithm 3.6).

We note that in Figure 4.8b the y-axis is $n\times$ MSE. The "CR bound" line is n times the trace of Cramér-Rao bound [Cramér, 1946, Rao, 1945], which is the asymptotic lower bound on the MSE of unbiased estimators. Because the Cramér-Rao bound decreases at the rate of $1/n$, the CR bound line is horizontal.

Observations.   Figure 4.8 shows the following order ("$\succ$" means "is better than"):

- Runtime: K-O $\succ$ Adaptive RBCML$\succ$ 1-LSR $\succ$ 2-LSR.

- Accuracy: 2-LSR $\succ$ Adaptive RBCML $\succ$ K-O$\succ$ 1-LSR. (Adaptive RBCML is slightly better than K-O.)

Thurstone's Case V Model.   Figure 4.9 shows the runtime and $n\times$MSE of two algorithms for Thurstone's Case V model for $m = 10$, where each utility distribution is a Gaussian distribution with variance 1. The synthetic experiment has 50,000 randomly generated preference profiles. In each preference profile, each ground truth parameter θ_i is generated uniformly at random from $[0, 5]$.

- **GMM$_{\text{RUM}}$** represents the GMM$_{\text{RUM}}$ algorithm for general RUMs (Algorithm 3.10).

- **RBCML($G_{\text{Full}}, \mathcal{W}_u$)** represents the one-round RBCML with unweighted full breaking and uniform CML weight graph. This is the only consistent RBCML method for the RUM according to Theorem 4.40.

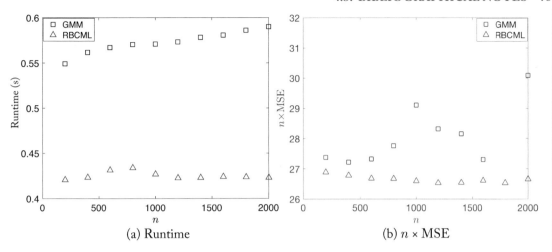

(a) Runtime

(b) $n \times$ MSE

Figure 4.9: The runtime and $n\times$MSE of GMM$_{\text{RUM}}$ and RBCML($G_{\text{Full}}, \mathcal{W}_u$) for RUM with Gaussian distributions.

Observations.   Figure 4.9 shows that the RBCML algorithm outperforms GMM$_{\text{RUM}}$ w.r.t. both runtime and MSE, and appears to be more stable in terms of MSE. Again, the y-axis in Figure 4.9b is $n\times$MSE.

4.3   BIBLIOGRAPHICAL NOTES

Section 4.1.   The section is largely based on Azari Soufiani et al. [2013a, 2014a].

Section 4.2.   The section is largely based on Zhao and Xia [2018]. The harmonic breaking (Example 4.19) and the K-O algorithm (Example 4.30) were introduced by Khetan and Oh [2016]. Khetan and Oh [2016] introduced and analyzed weighted breaking in a more general setting, where each agent submits a partial order with a special structure. They proved an upper bound on the worst-case sample complexity of any weighted breaking, and characterized the optimal weighted breaking w.r.t. the upper bound. The harmonic breaking is an example of their results applied to linear orders.

CHAPTER 5

Mixture Models for Rank Data

Recall that the goal of the group activity selection problem (Scenario 3 in Chapter 1) is to partition agents into groups and then recommend an activity for each group. This can be achieved by first clustering the agents according to their preferences over the activities, and then learn the common preferences for each group to recommend an activity.

One standard approach toward clustering is by using *mixture models* [McLachlan and Basford, 1988], where each cluster uses a (possibly different) statistical model to model the data generated from it. Additionally, mixture models often provide better fitness to data as shown in Table 2.1 for Preflib data.

Chapter Overview. In this chapter, we will see theories and algorithms for mixtures of a finite number of statistical models for rank data. Basic definitions, examples, and a generic EM algorithm will be presented in Section 5.1. Then, the identifiabilty and algorithms for mixtures of Plackett–Luce models, mixtures of general random utility models, and mixtures of Mallows' models will be discussed in Sections 5.2, Sections 5.3, and Sections 5.4, respectively.

5.1 MIXTURE MODELS

A mixture model is a statistical model that combines $K \geq 2$ statistical models $\mathcal{M}_1, \ldots, \mathcal{M}_K$ over the same sample space through a vector of *mixing coefficients* $\vec{\alpha} = (\alpha_1, \ldots, \alpha_K) \geq \vec{0}$ with $\vec{\alpha} \cdot \vec{1} = 1$, such that each data point is viewed as being generated from $\mathcal{M}_k$ with probability α_k.

Definition 5.1 Mixture Model. Given $K \geq 2$ statistical models $\{\mathcal{M}_k = (\mathcal{S}, \Theta_k, \vec{\pi}_k) : k \leq K\}$ over the same sample space $\mathcal{S}$, let $\mathrm{Mix}(\mathcal{M}_1, \ldots \mathcal{M}_K) = (\mathcal{S}, \Theta, \vec{\pi})$ denote the mixture model of $\mathcal{M}_1, \ldots, \mathcal{M}_K$, where

- **The sample space $\mathcal{S}$** is the same as that in $\mathcal{M}_1, \ldots, \mathcal{M}_K$.

- **The parameter space Θ** is $\Theta_1 \times \cdots \times \Theta_K \times \mathcal{C}_K$, where $\mathcal{C}_K \subseteq \mathbb{R}^K_{\geq 0}$ is the probability simplex that represents the mixing coefficients.

- **The probability distributions $\vec{\pi}$** consists of the following distributions: for any parameter $\vec{\theta} = (\vec{\theta}_1, \ldots, \vec{\theta}_K, \vec{\alpha}) \in \Theta$ and any $D \in \mathcal{S}$,

$$\pi_{\vec{\theta}}(D) = \sum_{k=1}^{K} \alpha_k \cdot \pi_{\vec{\theta}_k}(D).$$

This chapter will focus on a notable special case of mixture models, where all K models are the same. For any model $\mathcal{M}$, the mixture of K models, each of which being $\mathcal{M}$, is called "K-$\mathcal{M}$," and is denoted by $\mathrm{Mix}_K(\mathcal{M})$. Note that this does not mean that the mixing coefficients for the K components are identical or the parameters of the K components are identical.

Example 5.2 K-Plackett–Luce with The $\vec{\gamma}$ Parameterization.   Let $\mathcal{A} = \{a, b, c\}$. Figure 5.1 shows the probability distribution over $\mathcal{L}(\mathcal{A})$ in 2-Plackett–Luce, or 2-PL for short, given $(\vec{\gamma}_1, \vec{\gamma}_2, \vec{\alpha}) = ((0.1, 0.4, 0.5), (0.3, 0.3, 0.4), (0.3, 0.7))$, where $\vec{\gamma}_1$ is the parameter of the first PL component, $\vec{\gamma}_2$ is the parameter of the second PL component, and $(0.3, 0.7)$ are the mixing coefficients.

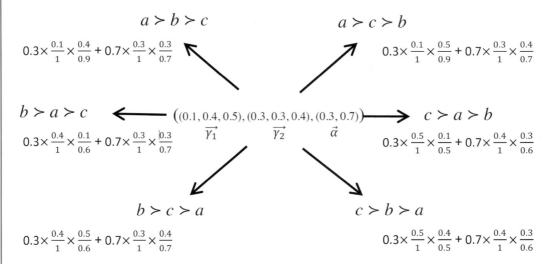

Figure 5.1: $\pi_{(\vec{\gamma}_1, \vec{\gamma}_2, \vec{\alpha})}$ in 2-PL, where $\vec{\gamma}_1 = (0.1, 0.4, 0.5)$, $\vec{\gamma}_2 = (0.3, 0.3, 0.4)$, and $\vec{\alpha} = (0.3, 0.7)$.

Using Mixture Models for Clustering.   Suppose the parameter $(\vec{\theta}_1, \ldots \vec{\theta}_K, \vec{\alpha})$ of a mixture model is learned from data D. Intuitively, this means that the population can be divided into K clusters, where for each $k \le K$, the k-th cluster is characterized by $\mathcal{M}_k$ with parameter $\vec{\theta}_k$ and size $\alpha_k \cdot n$. For each agent whose preferences are represented by $R \in D$, there are two ways to determine her membership in the K clusters.

- **Fractional membership.** For any $k \le K$, $\dfrac{\pi_{\vec{\theta}_k}(R)}{\sum_{l=1}^{K} \pi_{\vec{\theta}_l}(R)}$ portion of the agent belongs to cluster k.

- **Exclusive membership.** The agent exclusively belongs to cluster $k^* = \arg\max_k \pi_{\vec{\theta}_k}(R)$.

Example 5.3 Clustering by 4-PL on an Irish Election Poll Data [Gormley and Murphy, 2008]. Figure 5.2 shows the learned parameter of 4-PL from a 1997 Irish presidential election exit poll data [Gormley and Murphy, 2008].

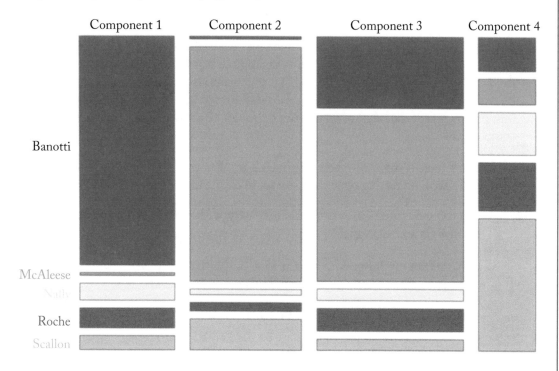

Figure 5.2: The learned parameter of 4-PL for five alternatives from a 1997 Irish presidential election exit poll data [Gormley and Murphy, 2008].

In Figure 5.2, there are five presidential candidates, each of which is represented by a color. There are four columns, each of which represents a PL component. The width of each component represents the mixing coefficient. Within each component, the height of each color is proportional to its γ value in Plackett–Luce.

How do we know the clustering results by mixture models, or more generally, the learned parameter of any statistical model, is correct? This question is often impossible to answer, because in many cases there is little or no information on the ground truth. That said, sometimes we can detect problematic cases by evaluating the identifiability (Definition 2.6) of a statistical model and the consistency of a parameter estimation algorithm (Definition 2.7). This chapter will focus on the identifiability of mixture models for rank data and consistency of algorithms for learning these mixture models.

5.1.1   IDENTIFIABILITY OF MIXTURE MODELS

Recall that the identifiability of a statistical model requires that different parameters lead to different distributions over the sample space. The identifiability of mixture models, such as K-PL, is more complicated, and there are two types of non-identifiability.

- The first type is *label switching*, which means that if we label the components of a mixture model differently, the distribution over samples does not change. For example, in 2-PL, the distribution over $\mathcal{L}(\mathcal{A})$ given $(\vec{\gamma}_1, \vec{\gamma}_2, (\alpha_1, \alpha_2))$ is the same as the distribution over $\mathcal{L}(\mathcal{A})$ given $(\vec{\gamma}_2, \vec{\gamma}_1, (\alpha_2, \alpha_1))$. Label switching can be addressed by introducing equivalent classes over parameters in the mixture model, or equivalently by ordering and merging duplicate components.

- The second type is more fundamental, which implies that the mixture model is non-identifiable even after using equivalent classes to eliminate the label switching problem.

In this chapter, (non-)identifiability refers to the second type. The next example shows that 2-PL for three alternatives is non-identifiable.

Example 5.4   Non-Identifiability of 2-PL for $m = 3$.   Let $\mathcal{A} = \{a, b, c\}$. Under 2-PL, the distribution over $\mathcal{L}(\mathcal{A})$ are the same given the following two parameters:

- $\vec{\theta} = ((0.8, 0.1, 0.1), (0.4, 0.3, 0.3), (0.3, 0.7))$ and

- $\vec{\eta} = ((0.6, 0.2, 0.2), (0.2, 0.4, 0.4), (0.8, 0.2))$.

To verify $\pi_{\vec{\theta}} = \pi_{\vec{\eta}}$, notice that $\gamma_b = \gamma_c$ in each of the four PL components (two from each 2-PL). Therefore, due to the symmetry of b and c, it suffices to check $\pi_{\vec{\theta}}(a \succ b \succ c) = \pi_{\vec{\eta}}(a \succ b \succ c)$ and $\pi_{\vec{\theta}}(b \succ a \succ c) = \pi_{\vec{\eta}}(b \succ a \succ c)$. It is not hard to verify that the former is 0.26 and the latter is $\frac{44}{300}$. Notice that the parameters for the four PL components are all different. This means that 2-PL for $m = 3$ is non-identifiable.

5.1.2   AN EM ALGORITHM FOR LEARNING MIXTURE MODELS

Recall that a mixture model assumes that each data point is generated from one of its components $\{\mathcal{M}_1, \ldots, \mathcal{M}_K\}$. If such membership information is available, then parameter estimation becomes much simpler—the mixing coefficient of $\mathcal{M}_k$ is proportional to the total weight of its members, and the parameter of $\mathcal{M}_k$ can be estimated from the preferences of its members.

The EM algorithm for mixture model (Algorithm 5.15) views memberships of the linear orders as missing data. More precisely, for any agent $j \leq n$ and any component $k \leq K$, let $z_{jk} \in \{0, 1\}$ be the latent membership variable, such that $z_{jk} = 1$ if R_j is generated from component $\mathcal{M}_k$; otherwise, $z_{jk} = 0$. Let $\vec{z}_j = (z_{j1}, \ldots, z_{jK})$ denote the latent membership variables for agent j, such that $\vec{z}_j \cdot \vec{1} = 1$. The structure of mixture models simplifies the E-step and the M-step in the EM algorithm shown in Algorithm 5.15; see McLachlan and Peel [2004, Section 2.8] for more details.

Algorithm 5.15 The EM Algorithm for Mixture Models [McLachlan and Peel, 2004].

Input: A preference profile D, a mixture model $\text{Mix}(\mathcal{M}_1, \ldots, \mathcal{M}_K)$, a number $T \in \mathbb{N}$.
Output: Estimated parameter of $\text{Mix}(\mathcal{M}_1, \ldots, \mathcal{M}_K)$.

1: Randomly initialize $\vec{\theta}^{(1)} = (\vec{\theta}_1^{(1)}, \ldots, \vec{\theta}_K^{(1)}, \vec{\alpha}^{(1)})$.
2: **for** $t = 1$ to T **do**
3: **E-Step:** For each $i \leq m$, each $j \leq n$, and each $k \leq K$, compute the expected membership of agent j in component k as follows:

$$\bar{z}_{jk}^{(t)} \propto \alpha_k^{(t)} \cdot \pi_{\vec{\theta}_k^{(t)}}(R_j).$$

For each $k \leq K$, compute the fractional profile for component k:

$$D_k^{(t)} = \{\bar{z}_{jk}^{(t)} @ R_j : j \leq n\}.$$

4: **M-Step:** For each $k \leq K$, compute $\vec{\theta}_k^{(t+1)}$ from D_k and let $\alpha_k^{(t+1)} = \sum_{j=1}^n \bar{z}_{jk}^{(t)}/n$.
5: **end for**
6: **return** $\vec{\theta}^{(T+1)}$.

The EM algorithm has a natural explanation. In the E-step, $\bar{z}_{jk}^{(t)}$ is computed to represent the expected membership of agent j in the k-th component $\mathcal{M}_k$. Note that unlike z_{jk}, $\bar{z}_{jk}^{(t)}$ can be a decimal. Then, R_j is divided into K fractional rankings, one for each component $\mathcal{M}_k$ with weight $\bar{z}_{jk}^{(t)}$. In the M-step, the parameter for the next round is calculated for each component $k \leq K$ from the fractional preference profile $D_k^{(t)}$ separately, and the mixing coefficient α_k is simply the fraction of agents belonging to $\mathcal{M}_k$.

We will see applications of Algorithm 5.15 later this chapter to mixtures of RUMs and mixtures of Mallows.

5.2   LEARNING MIXTURES OF PLACKETT–LUCE

Recall from Example 5.4 that 2-PL for three alternatives is not identifiable. This observation is generalized to K-PL for any $K \geq 2$ by the following theorem.

Theorem 5.5   Non-Identifiability of K-PL [Zhao et al., 2016].   *For any $m \geq 2$ and any $K \geq \frac{m+1}{2}$, K-PL for m alternatives is non-identifiable.*

Proof idea: The theorem is proved by explicitly constructing non-identifiable cases. In other words, a pair of different parameters, $(\vec{\gamma}_1, \ldots, \vec{\gamma}_K, \vec{\alpha})$ and $(\vec{\delta}_1, \ldots, \vec{\delta}_K, \vec{\beta})$, that lead to the same distribution over $\mathcal{L}(\mathcal{A})$ will be explicitly defined. Additionally, to simplify the construction, we let the values for $a_2, \ldots, a_m$ to be the same in each $\vec{\gamma}_k$ and $\vec{\delta}_k$.

This means that each $\vec{\gamma}_k = (\gamma_{k1}, \ldots, \gamma_{km})$ or $\vec{\delta}_k = (\delta_{k1}, \ldots, \delta_{km})$ is uniquely determined by its value for a_1. For any $k \leq K$, let $e_{2k-1} = \gamma_{k1}$ and let $e_{2k} = \delta_{k1}$. Let $\vec{\gamma}(e_1, e_3, \ldots, e_{2K-1}) = (\vec{\gamma}_1, \ldots, \vec{\gamma}_K)$ and $\vec{\delta}(e_2, e_4, \ldots, e_{2K}) = (\vec{\delta}_1, \ldots, \vec{\delta}_K)$.

The following stronger claim will be proved, which says that for any choice of $0 < e_1 < e_2 < \cdots < e_{2K}$, there exist $\vec{\alpha}$ and $\vec{\beta}$ so that $(\vec{\gamma}(e_1, e_3, \ldots, e_{2K-1}), \vec{\alpha})$ and $(\vec{\delta}(e_2, e_4, \ldots, e_{2K}), \vec{\beta})$ lead to the same distribution over $\mathcal{L}(\mathcal{A})$ under K-PL. The construction works as follows. For any $r \leq 2K$, define

$$\lambda_r = \frac{\prod_{p=1}^{2K-3}(pe_r + 2K - 2 - p)}{\prod_{q \neq r}(e_r - e_q)}.$$

Then, let $\vec{\alpha} \propto (\lambda_1, \lambda_3, \ldots, \lambda_{2K-1})$ and $\vec{\beta} \propto (\lambda_2, \lambda_4, \ldots, \lambda_{2K})$. In fact, Example 5.4 is an example of this construction for $K = 2$, $m = 3$, and $(e_1, e_2, e_3, e_4) = (0.2, 0.4, 0.6, 0.8)$. The full proof can be found in Zhao et al. [2016].   □

It follows from Theorem 5.5 that when a K-PL is used to cluster rankings over $m \leq 2K - 1$ alternatives, one must be careful about the learned parameter due to the non-identifiability of K-PL.

Example 5.6   In Example 5.3, 4-PL is applied to a dataset with 5 alternatives. According to Theorem 5.5, 4-PL for 5 alternatives is non-identifiable. Therefore, further verifications of the uniqueness of the learned parameter is needed.

The following theorem is a positive news: 2-PL is identifiable for any $m \geq 4$.

Theorem 5.7   Identifiability of 2-PL for $m \geq 4$ [Zhao et al., 2016].   *For any $m \geq 4$, 2-PL for m alternatives is identifiable.*

Proof idea: Let us sketch the proof for $m = 4$. It suffices to prove that for any different parameters of PL: $\vec{\gamma}_1, \vec{\gamma}_2, \vec{\gamma}_3, \vec{\gamma}_4$, the rank of the following $4! \times 4$ matrix is 4.

$$F = \begin{bmatrix} \pi_{\gamma_1}(V_1) & \pi_{\gamma_2}(V_1) & \pi_{\gamma_3}(V_1) & \pi_{\gamma_4}(V_1) \\ \pi_{\gamma_1}(V_2) & \pi_{\gamma_2}(V_2) & \pi_{\gamma_3}(V_2) & \pi_{\gamma_4}(V_2) \\ \vdots & \vdots & \vdots & \vdots \\ \pi_{\gamma_1}(V_{24}) & \pi_{\gamma_2}(V_{24}) & \pi_{\gamma_3}(V_{24}) & \pi_{\gamma_4}(V_{24}) \end{bmatrix}.$$

In F, $V_1, \ldots, V_{24}$ represents the 24 linear orders in $\mathcal{L}(\mathcal{A})$. The proof proceeds by constructing $4 \times 4!$ matrices H for different cases, such that $H \times F$ has full rank.

Let H_0 denote the matrix that extracts the probability for each alternative to be ranked at the top. Then,

$$H_0 \times F = \begin{bmatrix} 1 & 1 & 1 & 1 \\ \gamma_{11} & \gamma_{21} & \gamma_{31} & \gamma_{41} \\ \gamma_{12} & \gamma_{22} & \gamma_{32} & \gamma_{42} \\ \gamma_{13} & \gamma_{23} & \gamma_{33} & \gamma_{43} \end{bmatrix} = \begin{bmatrix} \vec{1} \\ \vec{\gamma}_{*1} \\ \vec{\gamma}_{*2} \\ \vec{\gamma}_{*3} \end{bmatrix},$$

where for each $l = 1, 2, 3$, $\vec{\gamma}_{*l} = [\gamma_{1l}, \gamma_{2l}, \gamma_{3l}, \gamma_{4l}]$.

If $\vec{1}, \vec{\gamma}_{*1}, \vec{\gamma}_{*2}$, and $\vec{\gamma}_{*3}$ are linearly independent, then the rank of F is 4, and the proof is done. Otherwise, suppose the rank of $H_0 \times F$ is 3 or less. W.l.o.g. suppose $\vec{1}$ and $\vec{\gamma}_{*1}$ be linearly independent, and suppose $\vec{\gamma}_{*2} = p \cdot \vec{\gamma}_{*1} + q$ for constants p and q with $p + q \neq 1$. Then, a matrix H_1 can be constructed such that:

$$H_1 \times F = \begin{bmatrix} 1 & 1 & 1 & 1 \\ \gamma_{11} & \gamma_{21} & \gamma_{31} & \gamma_{41} \\ \dfrac{1}{1 - \gamma_{11}} & \dfrac{1}{1 - \gamma_{21}} & \dfrac{1}{1 - \gamma_{31}} & \dfrac{1}{1 - \gamma_{41}} \\ \dfrac{1}{1 - p\gamma_{11} - q} & \dfrac{1}{1 - p\gamma_{21} - q} & \dfrac{1}{1 - p\gamma_{31} - q} & \dfrac{1}{1 - p\gamma_{41} - q} \end{bmatrix}.$$

If the rank of $H_1 \times F$ is 4, then the proof is done. Otherwise, if the rank of $H_1 \times F$ is 3 or less, then the four rows in $H_1 \times F$ are linearly dependent, which means that there exist four constants l_1, l_2, l_3, l_4 such that $\gamma_{11}, \gamma_{21}, \gamma_{31}$, and γ_{41} are four different roots to the following equation, because $\vec{\gamma}_k$'s are all different:

$$l_1 + l_2 \gamma + \frac{l_3}{1 - \gamma} + \frac{l_4}{1 - p\gamma - q} = 0$$
$$\Longleftrightarrow (l_1 + l_2 \gamma)(1 - p\gamma - q)(1 - \gamma) + l_3(1 - p\gamma - q) + l_4(1 - \gamma) = 0. \qquad (5.1)$$

However, due to the Fundamental Theorem of Algebra, Equation (5.1) has no more than three roots, which is a contradiction. Proof for other cases can be found in Zhao et al. [2016]. □

The next theorem is also positive, saying that under a mild condition, K-PL for m alternatives satisfies a weak notion of identifiability called *generic identifiability*, defined as follows.

Definition 5.8   Generic Identifiability of K-PL.   A parameter of a K-PL is *identifiable*, if there does not exist a different parameter that leads to the same distribution over $\mathcal{L}(\mathcal{A})$. A K-PL is *generically identifiable*, if the Lebesgue measure of identifiable parameters is 1.

Generic identifiability is desirable because it can be interpreted as "*model is generally sufficient for data analysis purposes*" [Allman et al., 2009].

Theorem 5.9   Generic Identifiability of K-PL [Zhao et al., 2016].   *For any $m \geq 6$ and any $1 \leq K \leq \lfloor \frac{m-2}{2} \rfloor!$, K-PL for m alternatives is generically identifiable.*

The proof is based on the uniqueness of tensor decomposition, which amounts to finding three independent random variables X, Y, and Z, given any parameter in $\mathcal{M}$ [Allman et al., 2009]. Let us recall the definition of tensors.

Definition 5.10　3-Way Rank-1 Tensor.　Given $\vec{x} \in \mathbb{R}^{d_1}$, $\vec{y} \in \mathbb{R}^{d_2}$, and $\vec{z} \in \mathbb{R}^{d_3}$, let $\vec{x} \otimes \vec{y} \otimes \vec{z}$ denote the 3-way rank-1 tensor, which is a $d_1 \times d_2 \times d_3$ array $\vec{P} = \{P_{l_1 l_2 l_3} \in \mathbb{R} : (l_1, l_2, l_3) \le (d_1, d_2, d_3)\}$, where $P_{l_1 l_2 l_3} = x_{l_1} \times y_{l_2} \times z_{l_3}$.

Tensor decomposition means writing a $d_1 \times d_2 \times d_3$ tensor Q as the linear combination of r rank-1 tensors, and the minimum r is called the *rank* of Q. That is,

$$Q = \vec{x}^{(1)} \otimes \vec{y}^{(1)} \otimes \vec{z}^{(1)} + \cdots + \vec{x}^{(r)} \otimes \vec{y}^{(r)} \otimes \vec{z}^{(r)}.$$

We now partition $\mathcal{A}$ to three subsets to represent K-PL as a three-way tensor. For convenience we will assume that m is even. The proof for odd m is similar:

$$A_X = \left\{ a_1, a_2, \cdots, a_{\frac{m-2}{2}} \right\}$$
$$A_Y = \left\{ a_{\frac{m}{2}}, a_{\frac{m+2}{2}}, \cdots, a_{m-2} \right\}$$
$$A_Z = \{a_{m-1}, a_m\}.$$

There are $\frac{m-2}{2}!$ rankings over A_X and A_Y, respectively, and two rankings over A_Z. Recall that a random variable is a mapping from the sample space to an outcome space. We now define three random variables that will be used in the tensor decomposition: (1) $X : \mathcal{L}(\mathcal{A}) \to \mathcal{L}(A_X)$, (2) $Y : \mathcal{L}(\mathcal{A}) \to \mathcal{L}(A_Y)$, and (3) $Z : \mathcal{L}(\mathcal{A}) \to \mathcal{L}(A_Z)$.

Definition 5.11　For any $R \in \mathcal{L}(\mathcal{A})$, let X, Y, and Z be three random variables that represent the restriction of a given linear order to A_X, A_Y, and A_Z, respectively. For any $R \in \mathcal{L}(\mathcal{A})$, let $\vec{x}(R)$ denote the $\frac{m-2}{2}!$ array, where the components are indexed by $\mathcal{L}(A_X)$, and the $X(R)$ component of $\vec{x}(R)$ is 1 and all other elements are 0's. $\vec{y}(R)$ and $\vec{z}(R)$ are defined similarly.

Because A_X, A_Y, and A_Z are non-overlapping, by Property 2.17, given any parameter $\vec{\gamma}$ of PL, X, Y, and Z are independent w.r.t. $\pi_{\vec{\gamma}}$. Therefore, given any parameter $\vec{\theta}$ of K-PL, let $\pi_{(X,Y,Z)}$ denote the joint probability distribution of X, Y, Z, represented by a $(\frac{m-2}{2}!) \times (\frac{m-2}{2}!) \times 2$ tensor, we have:

$$\pi_{(X,Y,Z)} = \sum_{k=1}^{K} \alpha_k \cdot \mathbb{E}_{R \sim \pi_{\vec{\gamma}_k}} \left(\vec{x}(R) \otimes \vec{y}(R) \otimes \vec{z}(R) \right)$$
$$= \sum_{k=1}^{K} \alpha_k \cdot \vec{x}^{(k)}(\vec{\gamma}_k) \otimes y^{(k)}(\vec{\gamma}_k) \otimes z^{(k)}(\vec{\gamma}_k),$$

where $\vec{x}^{(k)}(\vec{\theta}_k) = \mathbb{E}_{R \sim \pi_{\vec{\theta}_k}}(\vec{x}(R))$, and $\vec{y}^{(k)}(\vec{\theta}_k)$ and $\vec{z}^{(k)}(\vec{\theta}_k)$ are defined similarly. See the upper part of Figure 5.3 for an illustration.

Given X, Y, Z, and $\pi_{(X,Y,Z)}$, we are ready to prove Theorem 5.9.

Proof sketch for Theorem 5.9. Theorem 5.9 is proved by verifying the sufficient condition for unique tensor decompositions provided by Kruskal [1977, Theorem 4a]. To this end, the *Kruskal's rank* will be analyzed for three matrices, defined as follows:

$$\mathbf{P}_X = \begin{bmatrix} \vec{x}^{(1)}(\vec{\gamma}_1) \\ \vdots \\ \vec{x}^{(K)}(\vec{\gamma}_K) \end{bmatrix}, \quad \mathbf{P}_Y = \begin{bmatrix} \vec{y}^{(1)}(\vec{\gamma}_1) \\ \vdots \\ \vec{y}^{(K)}(\vec{\gamma}_K) \end{bmatrix}, \quad \mathbf{P}_Z = \begin{bmatrix} \vec{z}^{(1)}(\vec{\gamma}_1) \\ \vdots \\ \vec{z}^{(K)}(\vec{\gamma}_K) \end{bmatrix}.$$

See Figure 5.3 for an illustration.

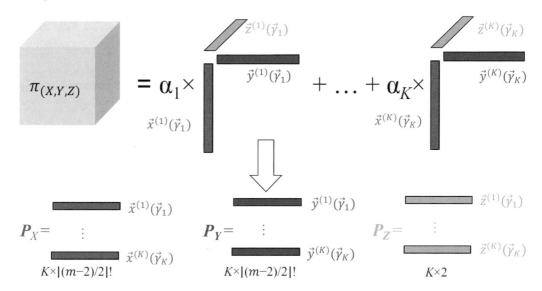

Figure 5.3: $\mathbf{P}_X$, $\mathbf{P}_Y$, and $\mathbf{P}_Z$ for verifying Kruskal's condition.

The Kruskal's (row) rank of a matrix is the maximum number l s.t. any l row of the matrix are linearly independent. Specifically, if a matrix has full row rank, then its Kruskal's rank is equal to the number of rows. Let $\mathrm{rank}_K(\mathbf{P}_X)$ denote the Kruskal's rank of matrix $\mathbf{P}_X$. The main theorem proved by Kruskal [1977] implies that $\pi_{(X,Y,Z)}$ has a unique decomposition if

$$\mathrm{rank}_K(\mathbf{P}_X) + \mathrm{rank}_K(\mathbf{P}_Y) + \mathrm{rank}_K(\mathbf{P}_Z) \geq 2k + 2. \tag{5.2}$$

When $k \leq \lfloor \frac{m-2}{2} \rfloor!$, it can be proved that $\mathbf{P}_X$ and $\mathbf{P}_Y$ generically have full row rank, because, take $k = \lfloor \frac{m-2}{2} \rfloor!$ for example, the determinant of $\mathbf{P}_X$ (respectively, $\mathbf{P}_Y$) can be represented by a polynomial over another polynomial. This means that Equation (5.2) generically holds. It follows that generically, $\pi_{(X,Y,Z)}$ has a unique tensor decomposition.

The uniqueness of tensor decomposition for $\pi_{(X,Y,Z)}$ does not immediately imply (generic) identifiability of K-PL. All it says is that (1) the mixing coefficients $\vec{\alpha}$ is unique and (2) for each $k \leq K$, the ratio among components in $\vec{\gamma}$ restricted to A_X (respectively, A_Y, A_Z) are the same. The generic identifiability of K-PL can be proved by applying the same argument to another partitioning $A_X' = \{a_m, \ldots, a_{\frac{m}{2}+2}\}$, $A_Y' = \{a_{\frac{m}{2}+1}, \ldots, a_3\}$, and $A_Z' = \{a_2, a_1\}$.   □

5.2.1   ALGORITHMS FOR MIXTURES OF PLACKETT–LUCE

The EM algorithm (Algorithm 5.15) can be directly applied to learn mixtures of PL. Notice that any parameter estimation algorithm can be used in Step 4 to compute the parameter for the next round. For example, the EMM algorithm by Gormley and Murphy [2008] can be viewed as Algorithm 5.15 with an MM algorithm in Step 4, where the GMM algorithm (Algorithm 3.7) and the LSR (Algorithm 3.6) can also be used.

In the rest of this section, we will see a GMM algorithm (Algorithm 5.16) with the following $m(m+1)$ moment conditions from three categories.

Category 1: Top-1 moment conditions. There are m moment conditions $\{g_i : i \leq m\}$, one for each alternative to be ranked at the top. Let $p_i = \sum_{r=1}^{K} \alpha_r \gamma_{ri}$. For any $R \in \mathcal{L}(\mathcal{A})$ and any $i \leq m$, define

$$g_i(R, \vec{\gamma}) = \begin{cases} 1 - p_i & \text{if } a_i \text{ is ranked at the top of } R \\ -p_i & \text{otherwise.} \end{cases}$$

Category 2: Top-2 moment conditions. There are $m(m-1)$ moments $\{g_{i_1 i_2} : i_1 \neq i_2 \leq m\}$, one for each combination of top-2 alternatives. Let $p_{i_1 i_2} = \sum_{r=1}^{K} (\alpha_r \cdot \frac{\gamma_{ri_1}\gamma_{ri_2}}{1-\gamma_{r,i_1}})$. For any $R \in \mathcal{L}(\mathcal{A})$ and any $i_1 \neq i_2 \leq m$, define

$$g_{i_1 i_2}(R, \vec{\gamma}) = \begin{cases} 1 - p_{i_1 i_2} & \text{if } a_i \text{ is ranked first and } a_{i_2} \text{ is ranked at the second in } R \\ -p_{i_1 i_2} & \text{otherwise.} \end{cases}$$

Category 3: Top-3 moment conditions. There are m moments that correspond to the following events represented by partial orders $[a_1 \succ a_2 \succ a_3 \succ \text{others}]$, $[a_2 \succ a_3 \succ a_4 \succ \text{others}]$, $\ldots$, $[a_{m-1} \succ a_m \succ a_1 \succ \text{others}]$, $[a_m \succ a_1 \succ a_2 \succ \text{others}]$. The corresponding $g_{i_1 i_2 i_3}$'s are defined similarly.

The following example shows the three categories of moment conditions. Let $K = 2, m = 4$, and let D consist of the following 20 rankings:

$$D = \{8@[a_1 \succ a_2 \succ a_3 \succ a_4], 4@[a_2 \succ a_3 \succ a_4 \succ a_1], 3@[a_3 \succ a_4 \succ a_1 \succ a_2],$$
$$3@[a_4 \succ a_1 \succ a_2 \succ a_3], 2@[a_3 \succ a_1 \succ a_4 \succ a_2]\}.$$

Category 1 example: for a_1 to be ranked at the top, we have:

$$g_1\left(D, \vec{\theta}\right) = \frac{8}{20} - (\alpha_1 \gamma_{11} + \alpha_2 \gamma_{21}).$$

Algorithm 5.16 GMM for 2-PL [Zhao et al., 2016] .

Input: A preference profile D.

Output: Estimated parameter of 2-PL.

  1: Compute the category 1, 2, 3 moment conditions given D.

  2: **return** the output of (3.6).

Category 2 example: for a_3 and a_4 to be ranked first and second, respectively, we have:

$$g_{34}\left(D, \vec{\theta}\right) = \frac{3}{20} - \left(\frac{\alpha_1 \gamma_{13} \gamma_{14}}{\gamma_{11} + \gamma_{12} + \gamma_{14}} + \frac{\alpha_2 \gamma_{23} \gamma_{24}}{\gamma_{21} + \gamma_{22} + \gamma_{24}}\right).$$

Category 3 example: for $[a_2 \succ a_3 \succ a_4 \succ a_1]$, we have

$$g_{234}\left(D, \vec{\theta}\right) = \frac{4}{20} - \left(\frac{\alpha_1 \gamma_{12} \gamma_{13} \gamma_{14}}{(\gamma_{13} + \gamma_{14} + \gamma_{11})(\gamma_{14} + \gamma_{11})} + \frac{\alpha_2 \gamma_{22} \gamma_{23} \gamma_{24}}{(\gamma_{23} + \gamma_{24} + \gamma_{21})(\gamma_{24} + \gamma_{21})}\right).$$

The three types of moment conditions are chosen based on the proof of Theorem 5.7 to guarantee that 2-PL is identifiable. The theoretical guarantee of Algorithm 5.16 is its consistency as shown in the following theorem.

Theorem 5.12   Consistency of Algorithm 5.16 [Zhao et al., 2016].   *Algorithm 5.16 is consistent if there exists $\epsilon > 0$ such that all parameters are in $[\epsilon, 1]$.*

The theorem is proved by verifying the sufficient conditions for GMM to be consistent [Hall, 2005], the ϵ requirement is used to guarantee the compactness of the parameter space.

5.3   LEARNING MIXTURES OF GENERAL RUMS WITH LOCATION FAMILIES

This section briefly discusses the identifiability and parameter estimation algorithms for mixtures of general RUMs with location families, in particular K-RUM.

Theorem 5.13   Non-Identifiability of RUMs with Location Families [Zhao et al., 2018b].
Let $RUM(\mathcal{M}_1, \ldots, \mathcal{M}_m)$ be an RUM with location families, where the utility distributions are symmetric. If $2K - 1 \geq m$, then $Mix_K(RUM(\mathcal{M}_1, \ldots, \mathcal{M}_m))$ is non-identifiable.

For example, the mixture of two Thurstone's Case V models over three alternatives is not identifiable. Note that Theorem 5.13 does not apply to mixtures of Plackett–Luce because the utility distributions are not symmetric.

Theorem 5.14 Generic Identifiability of RUMs with Location Families [Zhao et al., 2018b]. *Let $RUM(\mathcal{M}_1, \ldots, \mathcal{M}_m)$ be an RUM with location families, where all utility distributions have support $(-\infty, \infty)$. When $m \geq \max\{4K - 2, 6\}$, $Mix_K(RUM(\mathcal{M}_1, \ldots, \mathcal{M}_m))$ is generically identifiable.*

Like mixtures of Plackett–Luce models, mixtures of RUMs can be computed by the EM algorithm (Algorithm 5.15). For example, the EGMM algorithm by Zhao et al. [2018b] can be seen as Algorithm 5.15, where GMM_{RUM} (Algorithm 3.10) is used in step 4 to learn the parameters of each RUM component.

Next, we will introduce a GMM algorithm for K-RUM. In the GMM algorithm, the moment conditions are similar to the moment conditions for Algorithm 3.10. That is, for each pair of alternatives $a_{i_1} \neq a_{i_2}$, there is a moment condition $g_{i_1 i_2}$, defined as follows:

$$g_{i_1 i_2}\left(R, \vec{\theta}\right) = X^{a_{i_1} \succ a_{i_2}}(R) - \sum_{k=1}^{K} \alpha_k \cdot \pi_{\vec{\theta}_k}\left(a_{i_1} \succ a_{i_2}\right). \tag{5.3}$$

The following theorem says that Algorithm 5.17 is consistent for a natural class of mixture models. Like Algorithm 5.16, the theorem is proved by verifying the sufficient conditions by Hall [2005].

Algorithm 5.17 GMM for K-RUM [Zhao et al., 2018b].

Input: A preference profile D, a mixture model of K identical RUMs with location families.
Output: Estimated parameter.
  1: For all pairs of alternatives $a_{i_1} \neq a_{i_2}$, compute $X^{a_{i_1} \succ a_{i_2}}(D)$.
  2: Compute $\vec{\theta}^* \in \arg\min_{\vec{\theta}} \sum_{a \neq b} g_{i_1 i_2}(D, \vec{\theta})^2$ as defined in (5.3).
  3: **return** $\vec{\theta}^*$.

Theorem 5.15 Consistency of Algorithm 5.17 [Zhao et al., 2018b]. *Let $RUM(\mathcal{M}_1, \ldots, \mathcal{M}_K)$ be an RUM with location families. If $Mix_K(RUM(\mathcal{M}_1, \ldots, \mathcal{M}_K))$ is identifiable and the means of the utility distributions of all alternatives in all RUM components are bounded in $[0, C]$ for some constant C, then Algorithm 5.17 is consistent.*

5.4   LEARNING MIXTURES OF MALLOWS

This section briefly discusses the identifiability and algorithms for learning mixtures of Mallows' models with unfixed dispersion, where the dispersion value is also part of the parameter. Note that in other parts of this book, the dispersion value is fixed in Mallows' model (Definition 2.30).

Definition 5.16 Mallows' Model with Unfixed Dispersion.   Given $n \in \mathbb{N}$, the *Mallows' model with unfixed dispersion* is denoted by $\mathcal{M}_{\mathrm{Ma}} = (\mathcal{L}(\mathcal{A})^n, (\mathcal{L}(\mathcal{A}), \varphi), \vec{\pi})$, where $0 < \varphi < 1$, and for any $V, W \in \mathcal{L}(\mathcal{A})$, $\pi_{W,\varphi}(V) = \frac{1}{Z_{m,\varphi}} \varphi^{\mathrm{KT}(V,W)}$, where $Z_{m,\varphi} = \frac{\prod_{i=2}^{m}(1-\varphi^m)}{(1-\varphi)^{m-1}}$ is the normalization factor.

Again, the EM algorithm (Algorithm 5.15) can be applied to learn mixtures of Mallows. The identifiability of 2-Mallows with unfixed dispersion is implicitly proved by the algorithm by Awasthi et al. [2014], which will be presented in the rest of this section.

The algorithm by Awasthi et al. [2014] is based on tensor decomposition with the following reparameterization of Mallows.

Definition 5.17   Reparameterization of Mallows.   Given a parameter (W, φ) in a Mallows' model with unfixed dispersion, let

$$\vec{\delta}_{W,\varphi} = \left(\frac{\varphi^{W^{-1}[a_1]-1}}{Z_{m,\varphi}}, \frac{\varphi^{W^{-1}[a_2]-1}}{Z_{m,\varphi}}, \dots, \frac{\varphi^{W^{-1}[a_m]-1}}{Z_{m,\varphi}} \right).$$

Recall that $W^{-1}[a_i]$ is the rank of a_i in W. By Property 2.32, for any $i \le m$, the i-th component of $\vec{\delta}_{W,\varphi}$ is the probability for a_i to be ranked at the top in a linear order R generated from $\pi_{W,\varphi}$. Next, a random variable F will be defined to apply tensor decomposition.

Definition 5.18   Given $\vec{\delta}_{W,\varphi}$ and a partition A_1, A_2, A_3 of $\mathcal{A}$, let $F : \mathcal{L}(\mathcal{A}) \to \{0,1\}^{A_1 \times A_2 \times A_3}$ be the random variable such that for any $R \in \mathcal{L}(\mathcal{A})$, if the top three alternatives in R are $\{c_1, c_2, c_3\}$ (in any order), where $c_1 \in A_1$, $c_2 \in A_2$, $c_3 \in A_3$, then the (c_1, c_2, c_3) component of $F(R)$ is 1 and all other components are 0's; if no such $\{c_1, c_2, c_3\}$ exist, then all entries in $F(R)$ are zeros.

Given a parameter $\vec{\theta} = ((W_1, \varphi_1), (W_2, \varphi_2), \vec{\alpha})$ of 2-Mallows, we have:

$$\mathbb{E}_{R \sim \pi_{\vec{\theta}}} F(R) = \alpha_1 \cdot \mathbb{E}_{R \sim \pi_{W_1, \varphi_1}} F(R) + \alpha_2 \cdot \mathbb{E}_{R \sim \pi_{W_2, \varphi_2}} F(R).$$

By Property 2.32, for any $(c_1, c_2, c_3) \in A_1 \times A_2 \times A_3$, the (c_1, c_2, c_3) component of $\mathbb{E}_{R \sim \pi_{W,\varphi}} F(R)$ is:

$$\frac{(1+\varphi)\left(1+\varphi+\varphi^2\right)}{N_{m-2} N_{m-1} N_m \varphi^6} \cdot \varphi^{i_1+i_2+i_3} = \frac{N_m^2 (1+\varphi)\left(1+\varphi+\varphi^2\right)}{N_{m-2} N_{m-1} \varphi^3} \cdot \frac{\varphi^{i_1-1}}{N_m} \cdot \frac{\varphi^{i_2-1}}{N_m} \cdot \frac{\varphi^{i_3-1}}{N_m},$$

where i_1, i_2, i_3 are the ranks of c_1, c_2, c_3 in W, respectively.

For any W and φ, and any $i \in \{1, 2, 3\}$, let $\vec{x}_i(W, \varphi)$ be a vector of $|A_i|$ numbers that are indexed by $c_i \in A_i$, such that the c_i component is $\frac{\varphi^{W^{-1}[c_i]-1}}{N_m}$. In other words, for each $i \in \{1, 2, 3\}$, $\vec{x}_i(W, \varphi)$ is the restriction of $\vec{\delta}_{W, \varphi}$ on A_i. We have:

$$\mathbb{E}_{R \sim \pi_{\vec{\theta}}} F(R) = \frac{N_m^2 (1 + \varphi)\left(1 + \varphi + \varphi^2\right)}{N_{m-2} N_{m-1} \varphi^3} \left[\alpha_1 \cdot \vec{x}_1\left(W_1, \varphi_1\right) \otimes \vec{x}_2\left(W_1, \varphi_1\right) \otimes \vec{x}_3\left(W_1, \varphi_1\right) \right.$$
$$\left. + \alpha_2 \cdot \vec{x}_1\left(W_2, \varphi_2\right) \otimes \vec{x}_2\left(W_2, \varphi_2\right) \otimes \vec{x}_3\left(W_2, \varphi_2\right)\right].$$

The main idea behind the algorithm by Awasthi et al. [2014] (Algorithm 5.18) is the following. Given $\mathbb{E}_{R \sim \pi_{W_1, \varphi_1}} F(R)$, tensor decomposition is applied to learn $\vec{\delta}_{W_1, \varphi_1}$ and $\vec{\delta}_{W_2, \varphi_2}$ with high probability. A fallback algorithm is used to handle exceptions. In the algorithm, $\mathbb{E}_{R \sim \pi_{W_1, \varphi_1}} F(R)$ is approximated by $\sum_{R \in D} F(R)/n$. Details of the algorithm can be found in Awasthi et al. [2014], where the authors also proved that the algorithm has polynomial sample complexity.

Algorithm 5.18 Algorithm for 2-Mallows [Awasthi et al., 2014].

Input: A preference profile D.

Output: Estimated parameter of 2-Mallows with unknown dispersion.

1: Compute $\vec{F}(D) = \sum_{R \in D} F(R)/n$.
2: Apply any tensor decomposition algorithm to compute

$$\vec{F}(D) = \vec{x}_1^{(1)} \otimes \vec{x}_2^{(1)} \otimes \vec{x}_3^{(1)} + \vec{x}_1^{(2)} \otimes \vec{x}_2^{(2)} \otimes \vec{x}_3^{(3)}.$$

3: Learn $\vec{\theta}$ from $\vec{x}_1^{(1)}, x_2^{(1)}, \vec{x}_3^{(1)}, \vec{x}_1^{(2)}, \vec{x}_2^{(2)}, \vec{x}_3^{(3)}$.
4: **return** $\vec{\theta}$.

5.5   BIBLIOGRAPHICAL NOTES

Section 5.1.   The book by McLachlan and Peel [2004] offers a comprehensive introduction to finite mixture models, including the EM algorithm (Algorithm 5.15) and methods for choosing the number of components in a mixture model [McLachlan and Peel, 2004, Chapter 6].

Section 5.2.   The section is largely based on Zhao et al. [2016]. Allman et al. [2009] introduced generic identifiability and the proof technique based on Kruskal's rank for mixture models. Oh and Shah [2014] studied identifiability and algorithms for learning mixtures of multinomial logit model from partial-order data. Chierichetti et al. [2018] proposed a polynomial-time algorithm for learning the mixture of two multinomial logit models with the uniform mixing coefficients. Mollica and Tardella [2016] proposed a Bayesian approach toward learning mixtures of Plackett–Luce.

Section 5.3. The section is largely based on Zhao et al. [2018b].

Section 5.4. The section is largely based on Awasthi et al. [2014]. Liu and Moitra [2018] proved that K-Mallows is efficiently learnable for $K \geq 2$. The EM algorithm (Algorithm 5.15) is a special case of the algorithm by Lu and Boutilier [2014], which learns mixtures of Mallows from partial orders. Theory and algorithms for tensor decomposition can be found in the survey by Kolda and Bader [2009]. Mixtures of Mallows' models have been applied in assortment optimization [Desir et al., 2016]. Sample complexity for learning mixtures of Mallows has been studied by Chierichetti et al. [2015].

CHAPTER 6

Bayesian Preference Elicitation

In the picture ranking task (Scenario 2 in Chapter 1), suppose you have collected the following data for ranking three pictures $\{a, b, c\}$:

$$D = \{100@[a \succ c], 100@[b \succ c], 90@[c \succ a], 90@[c \succ b], 100@[a \succ b], 102@[b \succ a]\}.$$

The weighted majority graph of D (Definition 3.22) is shown in Figure 6.1. Suppose the price for an online worker to compare two pictures is $0.1, the price for an online worker to rank-order all three pictures is $0.2, and your remaining budget is $10. What is the best next question to ask?

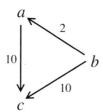

Figure 6.1: Weighted majority graph.

A straightforward answer is a vs. b, because they seem to be the most uncertain pair, and a linear order over all three pictures does not seem to provide much more information but costs twice as much as a pairwise comparison.

In many situations, however, the answer is not as clear. For example, what would be the answer if the number of $[c \succ a]$ in D were 102? What would be the answer if the goal were to choose top 2 pictures? What would be the answer if the remaining budget were $1,000?

More generally, the key question this chapter tries to answer is:

How can we elicit the most informative data under budget constraints?

Chapter Overview. In this chapter, we will discuss a principled Bayesian preference elicitation framework for answering this question. Section 6.1 introduces the Bayesian preference elicitation problem and a generic framework. Section 6.2 discusses some commonly used information criteria and approximation techniques for efficiently applying the framework to make a personal choice for a single agent. Section 6.3 focuses on making a social choice for a group of agents.

6.1   THE BAYESIAN PREFERENCE ELICITATION PROBLEM

The Bayesian preference elicitation problem consists of the following components.

- A set of agents and a set of alternatives $\mathcal{A}$. Each agent is characterized by a K-dimensional row vector $\vec{x}$ and each alternative is characterized by an L-dimensional row vector $\vec{z}$.

- A set of *designs* $\mathcal{H}$,[1] where each design $h \in \mathcal{H}$ is a query chosen by the decision-maker. For any $h \in \mathcal{H}$, let Answer(h) $\subseteq$ PO($\mathcal{A}$) denote all possible answers to h, and let Cost(h) denote the cost of h. After h is chosen by the decision-maker, an answer from Answer(h) is obtained as a new data point.

- A function $\mathcal{Q} : \bigcup_{k=1}^{\infty} \mathrm{PO}(\mathcal{A})^k \to \mathbb{R}$ that evaluates the quality of collected data.

Recall that PO($\mathcal{A}$) is the set of partial orders over $\mathcal{A}$. In this chapter, a design is a query about a chosen agent's preferences represented by a partial order, which means that the collected data at any point of time consist of partial orders.

Given data D that has been collected so far and the remaining budget W, the decision-maker's goal is to compute an optimal design to maximize the expected quality of data under a budget constraint. This process is illustrated in Figure 6.2 as a Markov decision process (MDP).

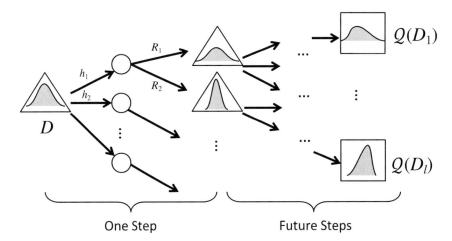

Figure 6.2: A Markov decision process (MDP) view of Bayesian preference elicitation.

In the MDP of Figure 6.2, each triangle is a decision node associated with the data collected so far. At a decision node, the decision-maker must choose a design, which is represented by an outgoing edge. Each circle is a chance node, whose outgoing edges represent possible

[1]The name follows after the convention in Bayesian experiment design [Chaloner and Verdinelli, 1995].

answers. There is a probability distribution associated with each chance node over its outgoing edges. The leaf nodes (squares) represent the dataset when the budget is depleted. The decision-maker obtains utility $Q(D')$ without discount when she reaches a leaf node D'.[2] The goal is to compute an optimal *policy*, which maps each decision node to an outgoing edge with maximum expected utility.

Unfortunately, the MDP cannot be easily solved by standard techniques due to the following reasons. First, the probability distributions at chance nodes are unknown. A statistical model is often assumed to capture such probabilities, and in this chapter we will use the Plackett–Luce model with features (Section 6.1.1). Second, the number of leaf nodes is exponentially large. This computational challenge is often tackled by various approximations as we will see later in this chapter.

Practically, the elicitation problem can be solved by the framework presented in Algorithm 6.19, where in each iteration, the decision-maker calls Algorithm 6.20 to compute the optimal design. Expectations in Steps 2 and 3 of Algorithm 6.20 are taken over the answers to h and answers to the optimal choices of designs in future steps.

Algorithm 6.19 Bayesian Preference Elicitation Framework.

Input: A Bayesian preference elicitation problem and a budget W.
Output: A decision made from collected data D.

  1: $D = \emptyset$.
  2: **while** W is enough for any design **do**
  3:     Compute an optimal design h^* by Algorithm 6.20.
  4:     The answer to h^* is added to D.
  5:     $W \leftarrow W - \text{Cost}(h^*)$.
  6: **end while**
  7: **return** a decision based on D.

Algorithm 6.20 Computing Optimal Design.

Input: Collected data D, a set of designs $\mathcal{H}$, and the remaining budget W.
Output: A design $h^* \in \mathcal{H}$.

  1: **for** each $h \in \mathcal{H}$ with $\text{Cost}(h) \leq W$ **do**
  2:     Compute the expected one-step information gain $\text{EIG}_1(h)$, e.g., by Algorithm 6.21.
  3:     Compute the expected future information gain $\text{EIG}_*(h)$.
  4: **end for**
  5: **return** $\arg\max_h (\text{EIG}_1(h) + \text{EIG}_*(h))$.

---

[2]If the decision-maker chooses to stop at a decision node with collected data D, then she obtains utility $Q(D)$.

6.1.1   PLACKETT–LUCE MODEL WITH FEATURES

We will use the Plackett–Luce model with features to model agents' preferences over alternatives, which will determine their answers to designs (queries). In the model, a parameter is denoted by $B = [b_{\kappa\iota}]_{K \times L}$, which is a matrix of real-valued coefficients that convert features of an agent and features of an alternative to a latent utility in the following way. Given B, agent j's latent utility for alternative a_i is:

$$u_{ji} = \vec{x}_j \, B \left(\vec{z}_i\right)^{\top}. \tag{6.1}$$

Therefore, $b_{\kappa\iota}$ corresponds to the correlation between the κ-th feature of an agent and the ι-th feature of an alternative. Then, for any agent j and any linear order $R_j = [a_{i_1} \succ a_{i_2} \succ \ldots \succ a_{i_m}]$, the probability of R_j is

$$\pi_B(R_j) = \frac{\exp(u_{ji_1})}{\sum_{q=1}^{m} \exp(u_{ji_q})} \times \frac{\exp(u_{ji_2})}{\sum_{q=2}^{m} \exp(u_{ji_q})} \times \cdots \times \frac{\exp(u_{ji_{m-1}})}{\exp(u_{ji_{m-1}}) + \exp(u_{ji_m})}. \tag{6.2}$$

For example, suppose suppose each movie is represented by L features, such as genre, production company, year of production, etc. Suppose each user is represented by K features, such as gender, age, career, etc. The parameter $b_{\kappa\iota}$ represents how much an agent with κ-th feature likes a movie with ι-th feature. Given a parameter B, an agent's non-deterministic preference over all movies is captured by Equation (6.2).

6.1.2   COMPUTING EXPECTED INFORMATION GAIN

This section discusses approximation techniques to efficiently compute EIG_1 and EIG_*. Let $\mathcal{H} = \{H_1, \ldots, H_q\}$. For any $h \in \mathcal{H}$, the answer to h is characterized by the Plackett–Luce model with features defined in Section 6.1.1. Following the literature in Bayesian experimental design [Chaloner and Verdinelli, 1995], for any design $h \in \mathcal{H}$, the *expected one-step information gain*, denoted by $\mathrm{EIG}_1(h)$, is computed in three stages illustrated in Figure 6.3 and Algorithm 6.21.

The expected future information gain $\mathrm{EIG}_*(h)$ is often hard to compute, and is often evaluated from $\mathrm{EIG}_1(h)$ by assuming that the one-step cost-effectiveness carries over to future steps. That is,

$$\frac{\mathrm{EIG}_*(h)}{W - \mathrm{Cost}(h)} = \frac{\mathrm{EIG}_1(h)}{\mathrm{Cost}(h)}.$$

The following sections will focus on two elicitation settings.

- **Personal Choice (Section 6.2).** Given a targeted agent with feature $\vec{x}_*$, the goal is to make a decision for her based on the prediction of her preferences.

- **Social Choice (Section 6.3).** Given a set of targeted agents with features $\{\vec{x}_1, \ldots, \vec{x}_n\}$, the goal is to predict the outcome of voting by them.

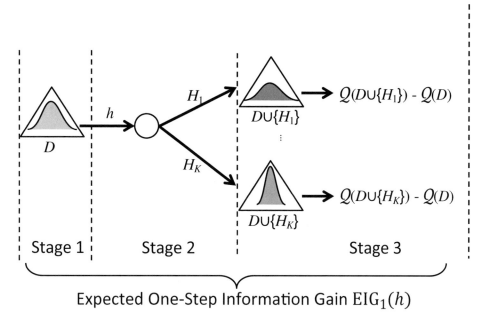

Figure 6.3: Computing the expected one-step information gain $\text{EIG}_1(h)$.

Algorithm 6.21 Expected One-Step Information Gain.

Input: Collected data D and a design $h \in \mathcal{H}$.
Output: $\text{EIG}_1(h)$.

1: Compute the distribution over the parameters given D, denoted by $\pi(\cdot|D)$.
2: For each answer $R \in \text{Answer}(h)$, compute its probability by

$$\Pr(R|D) = \int_{\Theta} \Pr(R|B)\pi(B|D)dB.$$

3: Compute the *expected information gain* for h:
$$\text{EIG}_1(h) = \mathbb{E}_R\left(\mathcal{Q}(D \cup \{R\}) - \mathcal{Q}(D)\right)$$
4: **return** $\text{EIG}_1(h)$.

In both settings, a budget W is given to create a dataset of preferences by eliciting preferences from other agents.

6.2 BAYESIAN PREFERENCE ELICITATION FOR PERSONAL CHOICE

In this section, we consider a simple preference elicitation setting for personal choice to illustrate the idea.

The Setting for Personal Choice. The preference elicitation setting for personal choice consists of the following components.

- Let $\vec{x}_*$ denote the features of the targeted agent.

- The set of designs $\mathcal{H}$ is a multiset of other agents' features.

- For each $h \in \mathcal{H}$, Answer$(h) = \mathcal{L}(\mathcal{A})$.

That is, a decision-maker chooses an agent (or equivalently, her feature) from $\mathcal{H}$ and asks her to report her linear order over $\mathcal{A}$. The collected linear orders will be used to estimate B to make a personal choices for the targeted agent.

Section 6.2.1 will introduce some commonly used information criteria. Section 6.2.2 discusses various approximation techniques to address the computational challenges in Algorithm 6.21. The elicitation algorithm will be formally presented in Section 6.2.2.

6.2.1 INFORMATION CRITERIA

One natural idea to evaluate the quality of data D is to measure the certainty in the posterior distribution over parameters given D. That is, $\mathcal{Q}(D) = \mathcal{I}(\pi(\cdot|D))$, where $\mathcal{I}$ is an information criterion that takes a distribution as input and outputs a real number that measures its certainty.

This leads to the following generic information criteria, which target at evaluating the "variance" of the posterior distribution in the Plackett–Luce model with features. For now, let us assume that the posterior distribution is a multivariate Gaussian distribution. The two criteria can be applied to arbitrary distributions.

Definition 6.1 D-Optimality [Mood et al., 1946, Wald, 1943] **and E-Optimality** [Ehrenfeld, 1955]**.** Given a Gaussian distribution $\mathcal{N}(\mu, \Sigma)$, the *D-optimality* measures the Shannon information gain, defined as $\mathcal{I}_D(\mu, \Sigma) = \det(\Sigma^{-1})$. The *E-optimality* is the minimum eigenvalue of the inverse of the covariance matrix, defined as $\mathcal{I}_E(\mu, \Sigma) = \lambda_{min}\{\Sigma^{-1}\}$.

It follows from the definition that the D-optimality and the E-optimality only involve the covariance matrix Σ, and larger values are favorable.

Generic information criteria may not be optimal for personal choice problems. For example, suppose the decision-maker's goal is to recommend one alternative to the targeted agent. In this case the targeted agent's preferences over other alternatives are irrelevant. To model various goals in personal choice, Azari Soufiani et al. [2013b] introduced a class of *minimum pairwise certainty (MPC)* criteria, defined as follows.

Definition 6.2   Minimum Pairwise Certainty (MPC) [Azari Soufiani et al., 2013b].   Given an m-dimensional Gaussian distribution $\mathcal{N}(\mu, \Sigma)$, where the i-th dimension u_i represents the targeted agent's utility for alternative a_i. The *pairwise certainty* for any pair of different alternatives a_{i_1}, a_{i_2} is defined as:

$$\text{PC}_{i_1 i_2}(\mathcal{N}(\mu, \Sigma)) = \frac{|\text{mean}(u_{i_1} - u_{i_2})|}{\text{std}(u_{i_1} - u_{i_2})},$$

where "mean" and "std" represent the mean and standard deviation of a distribution, respectively.

Given an unweighted graph $G = (\mathcal{A}, E)$ over $\mathcal{A}$, the *minimum pairwise certainty (MPC)* takes the minimum pairwise certainty value for all pairs of alternatives that are connected in G. That is,

$$\text{MPC}_G(\mathcal{N}(\mu, \Sigma)) = \min_{\{i_1, i_2\} \in E} \text{PC}_{i_1 i_2}(\mathcal{N}(\mu, \Sigma)).$$

Example 6.3   The following four types of MPC criteria are suitable for different personal choice scenarios.

(a) Full ranking is suitable for ranking the alternatives w.r.t. the distinguished agent's preferences.

(b) Ranked-top-k is suitable for recommending and ranking top k alternatives to the distinguished agent.

(c) Unranked-top-k is suitable for recommending k alternatives to the distinguished agent.

(d) Pairwise is suitable for comparing two alternatives.

Let $m = 6$ and suppose the ranking over alternatives is $[a_1 \succ \cdots \succ a_6]$ according to their means in $\mathcal{N}(\mu, \Sigma)$. Figure 6.4 shows the graphs for the four types of MPC.

6.2.2   APPROXIMATION TECHNIQUES FOR PERSONAL CHOICE

To address the challenge in computing $\text{EIG}_1(h)$ in Algorithm 6.21, the following approximation techniques are often used.

- **Approximation in Steps 1 and 2 of Algorithm 6.21.** A point estimation for B, e.g., maximum a posteriori (MAP) estimation, is used to calculate the distribution over answers.

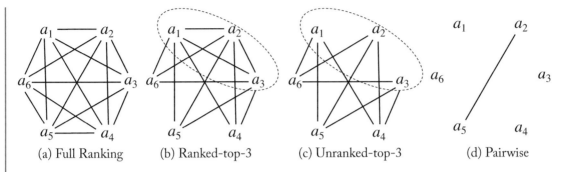

(a) Full Ranking (b) Ranked-top-3 (c) Unranked-top-3 (d) Pairwise

Figure 6.4: Four types MPC criteria.

- **Approximation in Step 3 of Algorithm** 6.21. The following Gaussian distribution is used to approximate the posterior distribution $\pi(\cdot|D \cup \{R\})$:

$$\mathcal{N}\left(\hat{B}, \left[P\left(\hat{B}\right) + J_{D\cup\{R\}}\left(\hat{B}\right)\right]^{-1}\right), \text{ where} \tag{6.3}$$

  - $\hat{B}$ is the MAP of $D \cup \{R\}$,

  - $P(\hat{B})$ is the *precision matrix* of the prior evaluated at $\hat{B}$. That is, $P(\hat{B}) = \nabla^2 \log \pi(\hat{B})$, where $\pi(\cdot)$ is the prior distribution, and

  - for any data D^*, $J_{D^*}(\hat{B})$ is the *observed information matrix*, defined as the negative Hessian of the log-likelihood, namely:

$$J_{D^*}\left(\hat{B}\right) = -\sum_{V \in D^*} \left(\nabla_B^2 \log \Pr(V|B)|_{B=\hat{B}}\right).$$

In the Gaussian approximation for Step 3, $P(\hat{B})$ characterizes the observed information in the prior distribution, and can be viewed as the posterior distribution over the parameters as if the prior distribution is uniform and there were an initial dataset D_0 obtained before D, such that for any pair of parameters B, B', we have $\frac{\Pr(D_0|B)}{\Pr(D_0|B')} = \frac{\pi(B)}{\pi(B')}$. Also notice that $J_{D^*}(\hat{B})$ is additive w.r.t. rankings in D^*. Therefore, when the data size is large, the approximation is dominated by the $J_{D^*}(\hat{B})$ component.

The Gaussian approximation also simplifies the calculation of information criteria in Definitions 6.1 and 6.2, because $u_{i_1} - u_{i_2}$ is a one-dimensional Gaussian distribution whose mean and variance can be easily computed by the mean and variance of their joint distribution.

Algorithm 6.22 is the refinement of Algorithm 6.20 for personal choice by adopting the approximation techniques discussed above. We recall that $\text{EIG}_*(h)$ is proportional to $\text{EIG}_1(h)$. Therefore, Algorithm 6.22 chooses the design h with maximum $\text{EIG}_1(h)$.

Algorithm 6.22 Computing Optimal Design for Personal Choice.

Input: Collected data D, a set of agents' features $\mathcal{H} \subseteq \mathbb{R}^K$, and a budget W.

Output: An optimal $h^* \in \mathcal{H}$.

1: **for** each $h \in \mathcal{H}$ with $\text{Cost}(h) \leq W$ **do**
2:　　Compute $\text{EIG}_1(h)$ by an information criterion in Section 6.2.1 using Gaussian approximation (6.3).
3: **end for**
4: **return** $h^* \in \arg\max_h \dfrac{\text{EIG}_1(h)}{\text{Cost}(h)}$.

6.3 BAYESIAN PREFERENCE ELICITATION FOR SOCIAL CHOICE

In this section, we consider a preference elicitation setting for a group of agents to make a social choice. Imagine that a hiring committee needs to make a decision on hiring one candidate by voting, but all committee members are unavailable at the moment to submit their preferences. Instead, we can elicit preferences of some carefully chosen agents who are not in the hiring committee, in order to learn the hiring committee members' preferences and predict the outcome of voting. How should we do it in an efficient and effective way?

The Setting for Social Choice.　The preference elicitation setting for social choice in this section consists of the following components.

- Social choice will be made for a group of distinguished agents, called the *key group*, whose features are denoted by a multiset $\mathcal{T}_{\text{key}}$. Other agents are called the *regular group*, whose features are denoted by a multiset $\mathcal{T}_{\text{reg}}$.

- Each design consists of two parts: $h = (\vec{x}, q)$, where $\vec{x} \in \mathcal{T}_{\text{reg}}$ is the feature of an agent in the regular group and q is a question about her linear order over $\mathcal{A}$. A general class of questions called *ranked-top-k* will be discussed in Section 6.3.2.

　　The elicitation framework for social choice (Algorithm 6.23) is illustrated in Figure 6.5, where the $1 \rightarrow 2 \rightarrow 3 \rightarrow 1$ loop is an extension of Algorithm 6.20 tailored for social choice.

6.3.1 APPROXIMATING POSTERIOR DISTRIBUTIONS

Recall that answers to a design are represented by partial orders over $\mathcal{A}$, which can be viewed as events in light of Section 2.1.1. Therefore, for any parameter B, $\pi(B|D)$ can be computed by Bayes' rule as follows:

$$\pi(B|D) = \frac{\Pr(D|B)\pi(B)}{\int_\Theta \Pr\left(D|\hat{B}\right) \pi\left(\hat{B}\right) d\hat{B}}.$$

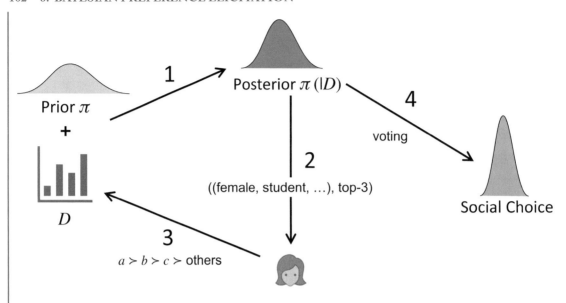

Figure 6.5: Illustration of the elicitation framework for social choice.

Algorithm 6.23 Elicitation Framework for Social Choice.

Input: A Bayesian preference elicitation problem for social choice and a budget W.
Output: A decision for the key group.

$D = \emptyset$.

while W is enough for any design **do**

    1. (Section 6.3.1) Compute $\pi(\cdot|D)$ based on collected data D and prior π.

    2. (Section 6.3.2) Compute the design $h^* = (\vec{x}, q)$ to maximize estimated expected information gain.

    3. A random agent with feature $\vec{x}$ is chosen from the regular group and is asked to answer question q; the answer is added to D; $W \leftarrow W - \text{Cost}(h^*)$.

end while

4. (Section 6.3.3) Compute the outcome of voting from D.

return the outcome of voting.

Like the Gaussian approximation for personal choice (Section 6.2.2), $\pi(B|D)$ can be approximated by a multivariate Gaussian distribution characterized by the composite marginal likelihood (Section 4.2.2) [Pauli et al., 2011].

For the purpose of presentation, let $\vec{\beta}$ be the $K \times L$ dimensional row vector that represents the vectorization of B. The composite marginal likelihood method [Lindsay, 1988] computes the estimate of the ground truth parameter from events, e.g., pairwise comparisons as in (4.5).

Given a dataset D of partial orders, let

$$\vec{\beta}_{\text{CML}} = \arg\max_{\vec{\beta} \in \Theta} \text{CLL}(\vec{\beta}, D) = \arg\max_{\vec{\beta} \in \Theta} \sum_{R \in D} \ln \pi_{\vec{\beta}}(R),$$

where $\text{CLL}(\vec{\beta}, D)$ is the composite log-likelihood function (4.5). Under the Plackett–Luce model with features, $\text{CLL}(\vec{\beta}, D)$ is twice differentiable for all $\vec{\beta} \in \Theta$, which means that $J(\vec{\beta}, D) = -\nabla^2_{\vec{\beta}} \text{CLL}(\vec{\beta}, D)$ is well defined.

Theorem 6.4 Gaussian Approximation of $\pi(\vec{\beta}|D)$ [Pauli et al., 2011]. *As $n \to \infty$, $\pi(\vec{\beta}|D)$ is approximately a multivariate Gaussian distribution $\mathcal{N}(\vec{\beta}_{\text{CML}}, J^{-1}(\vec{\beta}, D))$.*

This is the approximation used in Algorithm 6.23. Computing $J(\vec{\beta}, D)$ requires the computation of second-order partial derivatives of $\ln \pi_{\vec{\beta}}(D)$. We will see in the next section that this can be efficiently done under the Plackett–Luce model with features.

6.3.2 RANKED-TOP-k QUESTIONS

In this section, we will show how to derive $J(\vec{\beta})$ for *ranked-top-k* questions under the Plackett–Luce model with features.

Definition 6.5 Ranked-Top-k Questions. For any integers $k < l \le m$ and any $A_l \subseteq \mathcal{A}$ with $|A_l| = l$, let q_{k,A_l} denote the question that asks an agent to give a linear order over her top-k alternatives in A_l.

We note that ranked-top-k questions are different from the MPC ranked-top-k criterion (Example 6.3).

Example 6.6 A pairwise comparison is a ranked-top-k question, where $k = 1$ and $l = 2$. Choice data (C, c) (Section 2.2.4) is a ranked-top-k question, where $k = 1$, $l = |C|$, and $A_l = C$.

We note that $J(\vec{\beta}, D) = \sum_{R \in D} J(\vec{\beta}, R)$. Therefore, it suffices to show how to compute $J(\vec{\beta}, R)$ for a partial order R. Let R_j denote agent j's answer to a ranked-top-k question. W.l.o.g. let $A_l = \{a_1, \ldots, a_l\}$ and $R_j = [a_1 \succ a_2 \succ \ldots \succ a_k \succ (\text{others in } A_l)]$. Recall that given B, for any $i \le m$, $u_{ji} = \vec{x}_j B(\vec{z}_i)^\top$ represents agent j's latent utility for a_i. By Property 2.19, we have:

$$\pi_{\vec{\beta}}(R_j) = \prod_{p=1}^{k} \frac{\exp(u_{jp})}{\sum_{i=p}^{l} \exp(u_{ji})}, \text{ and}$$

$$\text{CLL}(\vec{\beta}, R_j) = \sum_{p=1}^{k} \left(u_{jp} - \ln \sum_{i=p}^{l} \exp(u_{ji}) \right).$$

For any $1 \leq \kappa \leq K$ and any $1 \leq \iota \leq L$, let $b_{\kappa\iota}$ denote the (κ, ι) entry of B. We have

$$\frac{\text{CLL}\left(\vec{\beta}, R_j\right)}{\partial b_{\kappa\iota}} = \sum_{p=1}^{k} \left(\frac{\partial u_{jp}}{\partial b_{\kappa\iota}} - \frac{\sum_{i=p}^{l} \exp\left(u_{ji}\right) \frac{\partial u_{ji}}{\partial b_{\kappa\iota}}}{\sum_{i=p}^{l} \exp\left(u_{jp}\right)} \right),$$

where $\frac{\partial u_{jp}}{\partial b_{\kappa\iota}}$ is a constant. In fact, $\frac{\partial u_{jp}}{\partial b_{\kappa\iota}}$ is the product of the κ-th feature of agent j and the ι-th feature of alternative a_p. Similarly, $\frac{\partial u_{ji}}{\partial b_{\kappa\iota}}$ is also a constant. Again, $\vec{\beta}$ is the vectorization of B. The diagonal entries $\nabla^2 \text{CLL}(\vec{\beta}, R_j)$ are:

$$\frac{\text{CLL}\left(\vec{\beta}, R_j\right)}{\partial b_{\kappa\iota}^2} = \sum_{p=1}^{k} \left(\left(\frac{\sum_{i=p}^{l} \exp\left(u_{ji}\right) \frac{\partial u_{ji}}{\partial b_{\kappa\iota}}}{\sum_{i=p}^{l} \exp\left(u_{ji}\right)} \right)^2 - \frac{\sum_{i=p}^{l} \exp\left(u_{ji}\right) \left(\frac{\partial u_{jp}}{\partial b_{kl}} \right)^2}{\sum_{p=1}^{l} \exp\left(u_{jp}\right)} \right).$$

For non-diagonal entries, we have

$$\frac{\text{CLL}\left(\vec{\beta}, R_j\right)}{\partial b_{\kappa_1\iota_1} \partial b_{\kappa_2\iota_2}} =$$

$$\sum_{p=1}^{k} \left(\frac{\left(\sum_{i=p}^{l} \exp\left(u_{ji}\right) \frac{\partial u_{ji}}{\partial b_{\kappa_1\iota_1}} \right) \left(\sum_{i=p}^{l} \exp\left(u_{ji}\right) \frac{\partial u_{ji}}{\partial b_{\kappa_2\iota_2}} \right)}{\left(\sum_{i=p}^{m'} e^{u_{ji}} \right)^2} - \frac{\sum_{i=p}^{l} \exp u_{ji} \left(\frac{\partial u_{ji}}{\partial b_{\kappa_1\iota_1}} \right) \left(\frac{\partial u_{ji}}{\partial b_{\kappa_2\iota_2}} \right)}{\sum_{i=p}^{l} \exp\left(u_{ji}\right)} \right).$$

Information Criteria.   With the Gaussian approximation of the posterior distribution in Theorem 6.4, D-optimality and E-optimality (Definition 6.1) can be readily applied. MPC (Definition 6.2) can be extended to cope with the key group by considering the MPC across all agents in the key group.

6.3.3   SOCIAL CHOICE BY RANDOMIZED VOTING

Given the preferences of agents in the key group, a *voting rule* can be applied to make a social choice. The challenge is, even in the ideal case where the parameter B of the Plackett–Luce model with features is learned with certainty, there is still randomness in the preferences of agents in the key group. In this section, we will see how to tackle this challenge by applying *randomized positional scoring rules*.

Definition 6.7   Randomized Positional Scoring Rules.   A randomized positional scoring rule $r_{\vec{s}}$ is characterized by a scoring vector $\vec{s} = (s_1, \ldots, s_m)$ with $s_1 \geq s_2 \geq \cdots \geq s_m$. For any preference profile D of n linear orders over $\mathcal{A}$ and any alternative $a \in \mathcal{A}$, let $\vec{s}(a, D) = \sum_{R \in D} s_{R^{-1}[a]}$ denote the total score of a in D. Then, alternative a is chosen as the winner with probability proportional to $\vec{s}(a, D)$.

For example, the randomized *plurality* rule is a randomized positional scoring rule with $\vec{s} = (1, 0, \ldots, 0)$. The randomized *Borda* rule is a randomized positional scoring rule with $\vec{s} = (m - 1, m - 2, \ldots, 0)$.

Example 6.8   Let $A = \{a_1, a_2, a_3\}$ and $D = \{[a_1 \succ a_2 \succ a_3], [a_1 \succ a_3 \succ a_2], [a_2 \succ a_1 \succ a_3]\}$. The plurality and Borda scores are shown in Table 6.1.

Table 6.1: The plurality scores and Borda scores in a preference profile

	a_1	a_2	a_3
Plurality	2	1	0
Borda	5	3	1

Under randomized plurality, a_1 wins with probability 2/3 and a_2 wins with probability 1/3. Under randomized Borda, a_1, a_2, and a_3 win with probabilities 5/9, 3/9, and 1/9, respectively.

Now, let us consider agent j's random preferences. Given a positional scoring rule $r_{\vec{s}}$, let X_{ji} denote the random variable that represents the score of a_i w.r.t. agent j's preferences. The following theorem shows that the probability for a_i to be the winner under $r_{\vec{s}}$ is proportional to the expected total score of a_i.

Theorem 6.9   [**Zhao et al., 2018a**].   *For any randomized positional scoring rule $r_{\vec{s}}$ and any alternative $a_i \in A$,*

$$\Pr(a_i \text{ wins under } r_{\vec{s}}) \propto \sum_{n}^{j=1} \mathbb{E}(X_{ji}).$$

Recall that under the Plackett–Luce model with features, given B, an agent's random preferences can be calculated by Equation (6.2). However, this does not immediately give us an efficient algorithm for computing $\mathbb{E}(X_{ji})$ in Theorem 6.9, because $\mathbb{E}(X_{ji})$ requires the computation of the probability for an alternative to be ranked at a given position, which is generally hard under Placket–Luce.

Fortunately, for randomized plurality and randomized Borda, $\mathbb{E}(X_{ji})$ can be efficiently computed based on the following observations:

- randomized plurality: $\mathbb{E}(X_{ji}) = \Pr(a_i \text{ is ranked at the top position in } R_j)$; and

- randomized Borda: $\mathbb{E}(X_{ji}) = \sum_{i' \neq i} \Pr(a_i \succ a_{i'} \text{ in } R_j)$.

Both probabilities have closed-form formulas under the Plackett–Luce with features. Following (6.2), the probability for a_{i_1} to be ranked at the top among $\{a_{i_1}, \ldots, a_{i_l}\}$ by agent j is

$\dfrac{\exp(u_{ji_1})}{\sum_{q=1}^{l} \exp(u_{ji_q})}$; for any pair of different alternatives a_1 and a_2, the probability of $a_1 \succ a_2$ by

agent j is $\dfrac{\exp(u_{j1})}{\exp(u_{j1}) + \exp(u_{j2})}$.

Therefore, for any given B, Theorem 6.9 can be efficiently applied to randomized plurality and randomized Borda. The expectation of the winning probabilities, denoted by $\bar{\pi}$, can also be efficiently computed given the Gaussian approximation of $\pi(B|D)$ (Theorem 6.4).

There are many ways that the expected winning probabilities $\bar{\pi}$ can be used to make a social choice for the key group. For example, $\bar{\pi}$ itself can be directly used as the social choice, as in the experiments in the next section; the winner can be the alternative a with the highest $\pi(a)$; or, any alternative a can be chosen as the winner with probability $\bar{\pi}(a)$.

6.4   EXPERIMENTAL RESULTS

In this section, we will see two experiments on the Bayesian elicitation framework for social choice (Algorithm 6.23) from Zhao et al. [2018a]. The first experiment (Section 6.4.1) estimates the cost of ranked-top-k questions on Amazon Mechanical Turk. The second experiment (Section 6.4.2) compares D-optimality, E-optimality, and MPC using synthetic data and the cost function learned from the first experiment.

6.4.1   ESTIMATING THE COST FUNCTION

Two types of experiments were run by Zhao et al. [2018a] on Amazon Mechanical Turk (MTurk) to estimate the cost function. Each task requires workers to answer a ranked-top-k question about their preferences over a set of hotels in New York City.

- **Type 1 experiments.** $k, l \in [2, 10], k = l - 1$. That is, a worker reports a linear order over l hotels.

- **Type 2 experiments.** $k \in [1, 10], l = 10$. That is, a worker reports a linear order over her top-k hotels from a set of 10 hotels.

Experiment Setting.   For Type 1 experiments, 54 Hotels in New York City are chosen to generate sets of l hotels to be shown to the MTurk workers. The hotels are anonymized and represented by 4 features: average guest rating on a travel website, price per night, time to Times Square, and time to the nearest airport. MTurk workers' features are not recorded.

For Type 2 experiments, sets of 10 hotels in New York City are randomly chosen and an MTurk worker is asked to rank her top-k hotels. An example of the user interface for ranking is show in Figure 6.6, where $k = 4$.

Experiment Results.   Figure 6.7 shows the average time a MTurk worker spent to submit a ranking, based on responses from 408 MTurk workers (202 for Type 1 experiments and 206 for Type 2 experiments).

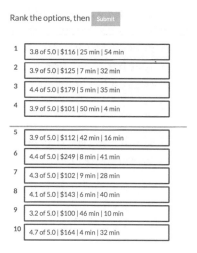

Figure 6.6: **Ranked-top-4.**

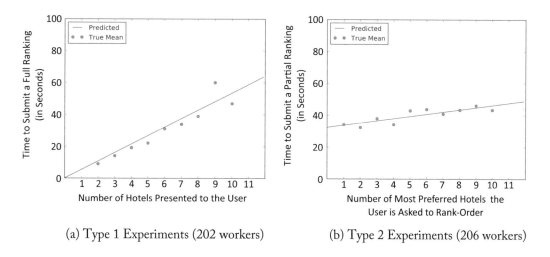

(a) Type 1 Experiments (202 workers)   (b) Type 2 Experiments (206 workers)

Figure 6.7: **Average time spent by an MTurk worker on ranked-top-k questions** [Zhao et al., 2018a].

It can be seen from Figure 6.7 that a linear cost function fits well. Based on the average hourly wage of MTurk workers ($3.16 USD per hour), the following cost functions are proposed.

- **Cost for Type 1 experiments.** The cost for ranking l alternatives is $0.0047l$.

- **Cost for Type 2 experiments.** The cost for ranking top-k alternatives from 10 alternatives is $0.0012k + 0.028$.

We note that the time a worker spent in Type 2 experiments is not very sensitive to k. This can be explained by the following three-step procedure: when a MTurk worker ranks her top-k choices, she may first read the descriptions of all hotels, then form her preferences about the hotels, and finally choose and rank-order her top-k hotels. The first and second steps do not depend on k and dominate the time spent on the last step.

6.4.2 COMPARING INFORMATION CRITERIA

The following synthetic data is used to test the Bayesian preference elicitation framework for social choice (Algorithm 6.23) with different information criteria [Zhao et al., 2018a].

Synthetic Data. Ten alternatives, 5 agents in the key group, and 20 agents in the regular group are randomly generated, each of which has 3 features. Each feature is independently generated from the standard Gaussian distribution $\mathcal{N}(0, 1)$. The ground truth parameter B is generated from a Dirichlet distribution $\mathrm{Dir}(\vec{1})$. The result is calculated for 400 trials. For each trial, the total budget is $0.9 and the cost functions learned from MTurk experiments (Section 6.4.1) are used.

Metrics. The total variation distance is used to measure the difference between the winning probabilities computed from the ground truth parameter and the winning probabilities computed by Algorithm 6.23, denoted by ψ^* and ψ, respectively.[3] The total variance distance is defined as:

$$\delta\left(\psi^*, \psi\right) = \frac{1}{2} \sum_{i=1}^{m} |\psi^*\left(a_i\right) - \psi\left(a_i\right)|.$$

Experiment Results. Figure 6.8 shows the total variance distance of four algorithms: Algorithm 6.23 with D-Optimality, E-Optimality, and MPC, respectively, and a naive algorithm that chooses a question under budget uniformly at random.

It can be seen from Figure 6.8 that for both randomized plurality and randomized Borda, the performances of MPC and D-optimality are better than E-optimality, and all of the three information criteria are better than the naive algorithm.

Figure 6.9 shows different types of questions asked by D-optimality and MPC in Algorithm 6.23 in different iterations. It can be seen that D-optimality almost always chooses a linear order as the most cost-effective question, while MPC tends to choose more linear orders than pairwise comparisons in early iterations. The distribution of types of questions for E-optimality is similar to that for MPC. Under all three information criteria, the top-1 question ($l = 10, k = 1$), which is equivalent to choice data, were rarely asked. This means that choice data is not as cost-effective as other types of data.

---

[3]While the accuracy of the algorithm w.r.t. learning the ground truth parameter is a natural measure, it is more important to measure how well the elicited preferences can help us make a better decision by estimating the distribution of the outcome of voting.

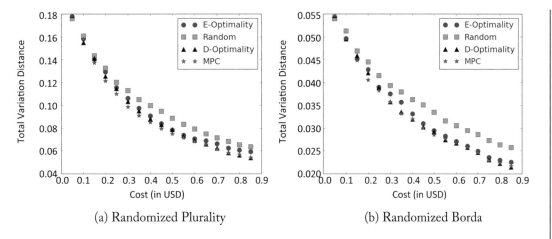

(a) Randomized Plurality   (b) Randomized Borda

Figure 6.8: Total variation distance of four elicitation algorithms [Zhao et al., 2018a].

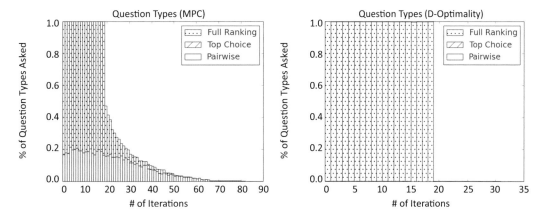

Figure 6.9: Types of questions asked by MPC (left) and by D-optimality (right) in different iterations of Algorithm 6.23 [Zhao et al., 2018a]. The legend "Full Ranking," "Top Choice," and "Pairwise" correspond to $(k = 9, l = 10)$, $(k = 1, l = 10)$, and $(k = 1, l = 2)$, respectively.

6.5 BIBLIOGRAPHICAL NOTES

Section 6.1. The framework of Bayesian preference elicitation is based on Bayesian experimental design [Chaloner and Verdinelli, 1995] and is closely related to active learning [Settles, 2012]. See Jiang et al. [2017] for some recent progresses on non-myopic active learning.

The preference elicitation problem has been widely studied in the field of recommender systems [Loepp et al., 2014], healthcare [Chajewska et al., 2000, Erdem and Campbell, 2017, Weernink et al., 2014], marketing [Huang and Luo, 2016], stable matching [Drummond and

Boutilier, 2014, Rastegari et al., 2016], combinatorial auctions [Sandholm and Boutilier, 2006], and social choice [Conitzer and Sandholm, 2005b]. There is also a line of research on using the minimax regret principle for preference elicitation [Boutilier, 2013].

Section 6.2.　The section is mostly based on Azari Soufiani et al. [2013b], which also includes algorithms for preference elicitation under general RUMs with features.

Sections 6.3 and 6.4.　The sections are mostly based on Zhao et al. [2018a]. Preference elicitation using Mallows' model was considered by Busa-Fekete et al. [2014].

CHAPTER 7

Socially Desirable Group Decision-Making from Rank Data

Socio-economic considerations are important in many decision-making scenarios that involve a group of humans. For example, in presidential elections (Scenario 1 in Chapter 1), fairness is a fundamental to democracy. In the group activity selection problem (Scenario 3 in Chapter 1), it is important to guarantee that the partition and activity selection are done in a fair way.

The challenge is, it is not always clear how such desiderata should be defined. For example, consider fairness in the product ranking problem (Scenario 6 in Chapter 1). A user may complain that the ranking is unfair to her, because her favorite product has received many awards but is ranked lower than expected by the website. See Figure 7.1 for an example on IMDB.

- **This is unfair! That film / show has received awards, great reviews, commendations and deserves a much higher vote! Can you change it to reflect this?**
  Our voting system is meant to offer a representation of what our users think of a film or show based on their votes. We do not collect or consider published reviews or critics' ratings or any other external factor: only votes cast by IMDb users are counted.
  We do not delete or alter individual votes and we do not adjust the results of our automated weighted rating for any individual movie / TV show. If the rating is lower than expected, it simply means that IMDb users who voted have a different opinion than those reviewers who liked it.

Figure 7.1: IMDb votes/ratings top frequently asked questions, March 2016.

As can be seen from Figure 7.1, it is impossible to guarantee that the outcome is always fair to everyone, simply because agents may have different, and sometimes conflicting preferences. Social choice theory offers one solution, which is to focus on the *procedure fairness* of voting rules for political elections.

For example, the *anonymity* axiom states that the voting rule is insensitive to permutations over agents' votes, which can be seen as a fairness criterion for the voters; *neutrality* is a fairness criterion for the alternatives; and *Condorcet criterion* (informally) states that an obviously socially strong alternative should win, which can be seen as a fairness criterion for strong alternatives. See Section 7.1.2 for mathematical definitions of some commonly studied fairness criteria.

This chapter will cover some preliminary approaches toward answering the following question.

How can we design an intelligent system for social choice (group decision-making) with desirable statistical properties, computational properties, and socio-economic properties such as fairness?

The question is easy to answer when there are two alternatives. In this case, the simple majority rule appears to be the best choice: it is the MLE of a simple yet natural statistical model [Condorcet, 1785], it satisfies many desirable axioms in social choice theory, and it is easy to compute.

Unfortunately, the question is hard to answer as soon as the number of alternatives is three, even w.r.t. just socio-economic criteria. Many impossibility theorems have been proved in social choice theory, saying that no mechanism can satisfy certain combinations of natural axioms. Therefore, tradeoffs must be made. The problem becomes even more challenging when statistical properties and computational properties are considered.

Chapter Overview.   Section 7.1 introduces a statistical decision-theoretic framework for social choice to consider statistical criteria, computational criteria, and socio-ecnomic criteria in the same mathematical framework. Section 7.2 provides a sufficient condition for the Bayesian estimator of a framework to satisfy a well studied frequentist measure called *minimaxity*. Section 7.3 explores the tradeoffs among satisfaction of the three types of criteria. In particular, we will see an impossibility theorem in Section 7.3.1, which states that no Bayesian estimator can satisfy a criterion called *strict Condorcet criterion*. On the positive side, Section 7.3.2 shows that many other socio-economic criteria, statistical criteria, and computational criteria can be satisfied at the same time. Section 7.4 briefly discusses some ideas on an automated framework for designing decision mechanisms with desirable properties.

7.1   STATISTICAL DECISION-THEORETIC FRAMEWORK

In previous chapters, we have seen various algorithms and procedures for parameter estimation. However, knowing the ground truth parameter does not always tell us how to make optimal decisions, as shown in the following example.

Example 7.1   Suppose Mallows' model is used to choose a single winner based on agents' preference profile D, as in Scenario 1 in Chapter 1. The estimated parameter, e.g., by MLE, cannot be directly used as the decision, because a parameter in Mallows' model is a linear order over $\mathcal{A}$, but the decision must be a single alternative.

There are two natural approaches in the social choice literature. Fishburn [1977] proposed to choose the top-ranked alternative in the MLE outcome as the winner.[1] Young [1988] proposed to choose the alternative with the highest probability to be ranked at the top, according to the posterior distribution over the parameters given D. Which approach should one follow?

---

[1]This voting rule is known as Kemeny (for winner), which is different from the Kemeny voting rule for choosing a linear order defined in Section 3.3.

Example 7.1 illustrates a critical gap between parameter estimation and decision-making. That is, the parameter space is sometimes different from the decision space. The gap is closed by *statistical decision theory* [Berger, 1985], where a loss function $L(\vec{\theta}, d)$ is introduced to evaluate the loss of a decision d w.r.t. the (unknown) ground truth parameter $\vec{\theta}$. In this chapter, we will use the following framework to evaluate decision-making mechanisms.

Definition 7.2 Statistical Decision-Theoretic Framework for Social Choice [Azari Soufiani et al., 2014b]. A *statistical decision-theoretic framework for social choice (framework* for short) is a tuple $\mathcal{F} = (\mathcal{M}, \mathcal{DS}, L)$, where $\mathcal{M} = (\mathcal{S}, \Theta, \vec{\pi})$ is a statistical model for rank data, $\mathcal{DS}$ is the decision space, and $L : \Theta \times \mathcal{DS} \to \mathbb{R}$ is a loss function.

In Example 7.1, the parameter space is $\Theta = \mathcal{L}(\mathcal{A})$ and the decision space is $\mathcal{DS} = \mathcal{A}$. The next example shows some natural combinations of decision spaces and loss functions.

Example 7.3 The 0-1 loss, top loss, and Borda loss are defined as follows.

- **0-1 loss.** Let $\mathcal{DS} = \Theta$. The 0-1 *loss function*, denoted by $L_{0\text{-}1}(\vec{\theta}, d)$, outputs 0 if $\vec{\theta}$ is identical to d; otherwise it outputs 1.

- **Top loss.** Let $\mathcal{DS} = \mathcal{A}$ and let Θ be $\mathcal{L}(\mathcal{A})$ or $\mathcal{B}(\mathcal{A})$. The *top loss function*, denoted by $L_{\text{top}}(\vec{\theta}, d)$, outputs 0 if for all $a \in \mathcal{A}$, $d \succeq a$ in $\vec{\theta}$; otherwise it outputs 1.

- **Borda loss.** Let $\mathcal{DS} = \mathcal{A}$ and let Θ be $\mathcal{L}(\mathcal{A})$ or $\mathcal{B}(\mathcal{A})$. The *Borda loss function*, denoted by $L_{\text{Borda}}(\vec{\theta}, d)$, outputs the number of alternatives that are preferred to d in $\vec{\theta}$, that is, $L_{\text{Borda}}(\vec{\theta}, d) = \#\{a \in \mathcal{A} : a \succ_{\vec{\theta}} d\}$.

All loss functions can be naturally generalized to evaluate a subset B of $\mathcal{DS}$ by computing the average loss of the decisions in B. More precisely, for any $B \subseteq \mathcal{DS}$ and any $\vec{\theta} \in \Theta$, let $L(\vec{\theta}, B) = \sum_{d \in B} L(\vec{\theta}, d)/|B|$.

The 0-1 loss function is often used to measure the quality of parameter estimation algorithms because the decision space is the same as the parameter space. The top loss and the Borda loss assume that the ground truth parameter is a linear order (as in Mallows' model) or a binary relation (as in Condorcet's model), and a decision is a co-winner. The loss is determined by the rank of the co-winner in the ground truth ranking.

7.1.1 MEASURING DECISION MECHANISMS: BAYESIAN LOSS AND FREQUENTIST LOSS

In practice, the ground truth parameter $\vec{\theta}$ is never known with certainty. Two principled approaches were pursued to deal with uncertainty in $\vec{\theta}$.

- **The Bayesian approach** models the parameter as part of the probability space, views $\pi_{\vec{\theta}}(D)$ as the conditional probability $\Pr(D|\vec{\theta})$, and requires a prior distribution π over the parameters.

- **The frequentist approach** assumes that the ground truth parameter is fixed but unknown, and the goal is to make a robust decision for all parameters.

The Bayesian approach and the frequentist approach differ not only in their philosophies in modeling uncertainty in the parameter, but also in their ways to measure decision mechanisms, which will be introduced soon. Let us start with defining decision mechanisms.

Definition 7.4 Decision Mechanisms.   Given a sample space S and a decision space DS, a decision mechanism $r : S \to (2^{DS} \setminus \emptyset)$ maps each data to a set of decisions.

The Bayesian (expected) loss for statistical decision-theoretic frameworks is defined as follows.

Definition 7.5 Bayesian Expected Loss.   Given a framework $\mathcal{F} = (\mathcal{M}, DS, L)$ and a prior distribution π over Θ, the *Bayesian expected loss* is a function $\mathrm{BL}_{\mathcal{F}} : DS \times S \to \mathbb{R}$, such that for any decision $d \in DS$ and any preference profile $D \in S$,

$$\mathrm{BL}_{\mathcal{F}}(d, D) = \int_{\Theta} L(\vec{\theta}, d) \Pr(\vec{\theta}|D) d\vec{\theta}.$$

That is, the Bayesian expected loss measures a decision d w.r.t. the data D by computing the expected loss of d w.r.t. the posterior distribution $\Pr(\cdot|D)$ over the parameters. It follows that the optimal decision mechanism in the Bayesian sense is the one that minimizes the Bayesian expected loss, and is called the *Bayesian estimator*.

Definition 7.6 Bayesian Estimator.   Given a framework $\mathcal{F} = (\mathcal{M}, DS, L)$ and a prior distribution π over Θ, the *Bayesian estimator* of $\mathcal{F}$, denoted by $\mathrm{BE}_{\mathcal{F}}$, is a decision mechanism such that for any preference profile D,

$$\mathrm{BE}_{\mathcal{F}}(D) = \arg\min_{d \in DS} \mathrm{BL}(d, D).$$

Let us see a few examples of Bayesian estimators.

Example 7.7   Let $\mathrm{BE}_{\mathrm{Ma},\varphi}^{\mathrm{Top}}$ denote the Bayesian estimator of the framework $(\mathcal{M}_{\mathrm{Ma},\varphi}, \mathcal{A}, L_{\mathrm{top}})$ with the uniform prior. That is, $\mathrm{BE}_{\mathrm{Ma},\varphi}^{\mathrm{Top}}$ is the optimal decision mechanism in the Bayesian sense under Mallows' model $\mathcal{M}_{\mathrm{Ma},\varphi}$ with the uniform prior, a decision is an alternative (co-winner), and the loss function measures whether the decision is ranked at the top of the ground truth linear order. This is Young [1988]'s approach mentioned in Example 7.1.

Let $\mathrm{BE}_{\mathrm{Co},\varphi}^{\mathrm{Borda}}$ denote the Bayesian estimator of the framework $(\mathcal{M}_{\mathrm{Co},\varphi}, \mathcal{A}, L_{\mathrm{Borda}})$ with uniform prior. That is, $\mathrm{BE}_{\mathrm{Co},\varphi}^{\mathrm{Borda}}$ is the optimal decision mechanism in the Bayesian sense under Condorcet's model $\mathcal{M}_{\mathrm{Co},\varphi}$ with the uniform prior, a decision is an alternative (co-winner), and the loss function measures the "rank" of the decision in the ground truth binary relation.

Now let us define the frequentist (expected) loss as follows.

Definition 7.8 Frequentist Expected Loss.   Given a framework $\mathcal{F} = (\mathcal{M}, \mathcal{DS}, L)$, a parameter $\vec{\theta} \in \Theta$, $n \in \mathbb{N}$, and a decision mechanism r, the *frequentist expected loss* $\mathrm{FL}_n(\vec{\theta}, r)$ is the expected loss of the output of r w.r.t. $\vec{\theta}$. More precisely,

$$\mathrm{FL}_n(\vec{\theta}, r) = \mathbb{E}_{D_n \in \mathcal{L}(A)^n \sim \pi_{\vec{\theta}}} L(\vec{\theta}, r(D_n)), \text{ where}$$

D_n is a randomly generated preference profile of n i.i.d. linear orders, each of which is generated from $\pi_{\vec{\theta}}$.

In other words, the frequentist expected loss evaluates a given decision mechanism r w.r.t. a given ground truth parameter $\vec{\theta}$, and the expectation is taken over randomly generated data D_n from $\mathcal{M}$ given $\vec{\theta}$. As frequentists do not assign likelihood to the ground truth parameters, a decision mechanism r is often measured by its worst-case loss w.r.t. all parameters $\vec{\theta}$. This leads to the definition of *minimax decision mechanisms* as follows.

Definition 7.9 Minimax Decision Mechanisms.   Given a framework $\mathcal{F} = (\mathcal{M}, \mathcal{DS}, L)$, a decision mechanism r is *minimax*, if for all $n \in \mathbb{N}$, $r \in \arg\min_{r'} \max_{\vec{\theta} \in \Theta} \mathrm{FL}_n(\theta, r')$.

That is, a minimax decision mechanism r minimizes the worst-case frequentist loss among all decision mechanisms. A minimax mechanism is desirable because it is the most robust mechanism against an adversarial nature who controls the true state of the world (the parameter).

A minimax decision mechanism can also be viewed as a mechanism with the optimal *sample complexity* in the framework where $\mathcal{DS} = \Theta$. Given a framework $\mathcal{F} = (\mathcal{M}, \Theta, L)$, a decision mechanism r, any $\epsilon > 0$, and any $\delta > 0$, the *sample complexity* of r is the minimum number of data points n such that

$$\forall \vec{\theta} \in \Theta, \quad \mathrm{Pr}_{D_n \sim \pi_{\vec{\theta}}} \left(L(\vec{\theta}, r(D_n)) > \epsilon \right) < \delta.$$

Let $L^*(\vec{\theta}, d) = \begin{cases} 0 & \text{if } L(\vec{\theta}, d) \leq \epsilon \\ 1 & \text{otherwise} \end{cases}$. The sample complexity of r is equivalent to the minimum n such that for all $\vec{\theta} \in \Theta$, $\mathrm{FL}_n(\vec{\theta}, r) < \delta$ in the framework $\mathcal{F}^* = (\mathcal{M}, \Theta, L^*)$. Therefore, any minimax decision mechanism of $\mathcal{F}^*$ has the lowest sample complexity in $\mathcal{F}$.

7.1.2   SOCIO-ECONOMIC CRITERIA: SOCIAL CHOICE AXIOMS

Recall from Definition 7.4 that a decision mechanism outputs a set of decisions for each preference profile. When $\mathcal{DS} = \mathcal{A}$, such mechanisms are called *irresolute rules* in social choice theory. Note that this means that all mechanisms are deterministic. This section covers some axioms defined in the social choice theory[2] for irresolute rules.

Definition 7.10   Anonymity.   An irresolute rule r satisfies *anonymity*, if for any preference profile $D = (R_1, \ldots, R_n)$ and any permutation σ_n over $\{1, \ldots, n\}$, we have

$$r(D) = r\left(R_{\sigma_n(1)}, \ldots, R_{\sigma_n(n)}\right).$$

In other words, r satisfies anonymity if the outcome is insensitive to the identities of the agents. Anonymity can be seen as a fairness condition for agents.

Definition 7.11   Neutrality.   An irresolute rule r satisfies *neutrality*, if for any preference profile $D = (R_1, \ldots, R_n)$ and any permutation $\sigma_{\mathcal{A}}$ over $\mathcal{A}$, we have

$$r\left(\sigma_{\mathcal{A}}(D)\right) = \sigma_{\mathcal{A}}(r(D)).$$

In other words, r satisfies neutrality if the outcome is insensitive to the identities of the alternatives. Neutrality can be seen as a fairness condition for alternatives.

The following three properties can be seen as fairness conditions for strong alternatives.

Definition 7.12   Condorcet Criterion.   An irresolute rule r satisfies *Condorcet criterion*, if for any preference profile D, whenever a *Condorcet winner* a exists, we must have $r(D) = \{a\}$. A Condorcet winner is an alternative that beats all other alternatives in their head-to-head competitions.

A Condorcet winner may not exist, but when it exists, it must be unique.

Definition 7.13   Strict Condorcet Criterion.   An irresolute rule r satisfies *strict Condorcet criterion*, if for any preference profile D, whenever the set of *weak Condorcet winners* is non-empty, it must be the outcome $r(D)$. A weak Condorcet winner is an alternative that never loses to any other alternatives in their head-to-head competitions.

When the number of agents is odd, strict Condorcet Criterion is equivalent to the Condorcet Criterion because there are no ties in head-to-head competitions. A weak Condorcet

---

[2]While mathematically they are just properties of mechanisms, we follow the social choice convention to call them "axioms."

winner may not exist, and when one exists, it may not be unique due to the ties among alternatives.

Definition 7.14 Monotonicity. An irresolute rule r satisfies *monotonicity*, if for any D, any $a \in r(D)$, and any D' that is obtained from D by only raising the position of a without changing the relative positions of other alternatives in each ranking, we have $a \in r(D')$.

That is, under a monotonic irresolute rule, if a is already winning in D, then in any profile D' where a is ranked higher, a should still be a co-winner.

7.2 MINIMAX ESTIMATORS IN NEUTRAL FRAMEWORKS

In this section, we will see a sufficient condition for the Bayesian estimator to satisfy minimaxity. Let us first define the neutrality of a framework for general decision spaces, which follows a similar idea behind the neutrality in Definition 7.11.

Intuitively, a framework $\mathcal{F} = (\mathcal{M}, \mathcal{DS}, L)$ is neutral if and only if all three components are neutral w.r.t. permutations over $\mathcal{A}$. Because the permutation over $\mathcal{A}$ may not be easily applicable to the parameter space and the decision space, we require the existence of homomorphisms from the permutation group over $\mathcal{A}$ to the permutation group over Θ and to the permutation group over $\mathcal{DS}$, respectively. Formally, the neutrality of a framework is defined as follows.

Definition 7.15 A framework $\mathcal{F} = (\mathcal{M}, \mathcal{DS}, L)$ where $\mathcal{M} = (\mathcal{S}, \Theta, \vec{\pi})$ is *neutral*, if each permutation σ over $\mathcal{A}$ can be mapped to a permutation σ_Θ over Θ and a permutation $\sigma_{\mathcal{DS}}$ over $\mathcal{DS}$ that satisfy the following conditions.

(i) Homomorphism. For any pair of permutations σ and σ' over $\mathcal{A}$, $(\sigma \circ \sigma')_\Theta = \sigma_\Theta \circ \sigma'_\Theta$ and $(\sigma \circ \sigma')_{\mathcal{DS}} = \sigma_{\mathcal{DS}} \circ \sigma'_{\mathcal{DS}}$.

(ii) Model neutrality. For any $\vec{\theta} \in \Theta$, any $R \in \mathcal{L}(\mathcal{A})$, and any permutation σ over $\mathcal{A}$, we have $\pi_{\vec{\theta}}(R) = \pi_{\sigma_\Theta(\vec{\theta})}(\sigma(R))$.

(iii) Loss function neutrality. For any $\vec{\theta} \in \Theta$, any $d \in \mathcal{DS}$, and any permutation σ over $\mathcal{A}$, we have $L(\vec{\theta}, d) = L(\sigma_\Theta(\vec{\theta}), \sigma_{\mathcal{DS}}(d))$.

In some cases the existence of σ_Θ and $\sigma_{\mathcal{DS}}$ is obvious as shown in the following example.

Example 7.16 For any $0 < \varphi < 1$, $(\mathcal{M}_{\text{Ma},\varphi}, \mathcal{A}, L_{\text{top}})$, $(\mathcal{M}_{\text{Ma},\varphi}, \mathcal{A}, L_{\text{Borda}})$, $(\mathcal{M}_{\text{Co},\varphi}, \mathcal{A}, L_{\text{top}})$, $(\mathcal{M}_{\text{Co},\varphi}, \mathcal{A}, L_{\text{Borda}})$ are neutral, where $\sigma_\Theta = \sigma_{\mathcal{DS}} = \sigma$.

The main theorem of this section says that if a neutral framework further satisfies the following connectivity condition, then its Bayesian estimator is a minimax mechanism.

(iv) Parameter connectivity. For any pair $\theta_1, \theta_2 \in \Theta$, there exists a permutation σ over $\mathcal{A}$ such that $\sigma_\Theta(\theta_1) = \theta_2$.

Theorem 7.17 [Xia, 2016]. *For any neutral framework $\mathcal{F}$ that satisfies parameter connectivity, $BE_{\mathcal{F}}$ with the uniform prior is minimax in $\mathcal{F}$.*

The theorem is proved by verifying that the uniform distribution over the parameter space is a *least favorable distribution*, by applying the following lemma.

Lemma 7.18   Section 5.3.2 III in Berger [1985]   *Given a framework $\mathcal{F}$, let r_{π^*} denote the Bayesian estimator for prior π^*. If $FL_n(\vec{\theta}, r_{\pi^*})$ are equal for all $\vec{\theta} \in \Theta$, then r_{π^*} is minimax.*

We note that Theorem 7.17 does not mix up the Bayesian principle with the frequentist principle. All it says is that a well-defined decision mechanism (BE$_{\mathcal{F}}$ with uniform prior) satisfies minimaxity.

It is not hard to verify that all models mentioned in Example 7.16 satisfy parameter connectivity. Therefore, their Bayesian estimators with the uniform prior are minimax. Notice that when the 0-1 loss function is used, the Bayesian estimator with the uniform prior becomes the MLE. Therefore, Theorem 7.17 immediately implies that the MLE is minimax.

Corollary 7.19   The MLE of Mallows' model is a minimax mechanism for $\mathcal{F} = (\mathcal{M}_{\mathrm{Ma},\varphi}, \mathcal{A}, L_{0\text{-}1})$.

The following example shows that parameter connectivity is necessary in Theorem 7.17. In other words, the Bayesian estimator of some neutral framework does not satisfy minimaxity.

Example 7.20   Let $\mathcal{A} = \{a, b\}$. Consider the following framework $(\mathcal{M}, \mathcal{A}, L)$, where $\mathcal{A} = \{a, b\}$, $\mathcal{M} = (\mathcal{L}(\mathcal{A})^n, \mathcal{L}(\mathcal{A}) \times \{0.6, 0.7\}, \vec{\pi})$. For any $(W, \varphi) \in \Theta$, $\pi_{W,\varphi}$ is equivalent to π_W in Mallows' model with dispersion φ. For any $W \in \mathcal{L}(\mathcal{A})$ and $a \in \mathcal{A}$, we let $L((W, 0.6), a) = L_{\mathrm{top}}(W, a)$ and $L((W, 0.7), a) = 1 - L_{\mathrm{top}}(W, a)$.

It can be verified that $\mathcal{F}$ is neutral by letting γ_Θ be the permutation that only applies to the first component of the parameter (the linear order). Let $n = 1$. When the preference profile is $\{a \succ b\}$ and the prior is uniform, the posterior distribution is shown in Table 7.1.

Table 7.1: The posterior distribution in the framework in Example 7.20

Parameter	$(a \succ b, 0.6)$	$(b \succ a, 0.6)$	$(a \succ b, 0.7)$	$(b \succ a, 0.7)$
Posterior Probability	$\dfrac{1}{3.2}$	$\dfrac{0.6}{3.2}$	$\dfrac{1}{3.4}$	$\dfrac{0.7}{3.4}$
Loss for a	0	1	1	0

Therefore, $\mathrm{BL}_{\mathcal{F}}(a, \{a \succ b\}) = \frac{0.6}{3.2} < \frac{0.7}{3.4} = \mathrm{BL}_{\mathcal{F}}(b, \{a \succ b\})$ which means that $\mathrm{BE}_{\mathcal{F}}(a \succ b) = a$. Similarly, $\mathrm{BE}_{\mathcal{F}}(b \succ a) = b$.

When the ground truth parameter is $(a \succ b, 0.7)$, the frequentist expected loss of $\mathrm{BE}_{\mathcal{F}}$ is $\frac{1}{1.7} > \frac{1}{2}$. We note that the worst-case frequentist loss of the decision mechanism that always outputs $\mathcal{A} = \{a, b\}$ is $\frac{1}{2}$, which means that $\mathrm{BE}_{\mathcal{F}}$ is not minimax.

7.3   SOCIALLY DESIRABLE BAYESIAN ESTIMATORS

In this section we will analyze the satisfaction of axioms in Section 7.1.2 for Bayesian estimators under frameworks where the decision space $\mathcal{S}$ is $\mathcal{A}$.

7.3.1   AN IMPOSSIBILITY THEOREM ON STRICT CONDORCET CRITERION

Theorem 7.21   [Xia, 2016].   *For any framework $\mathcal{F} = (\mathcal{M}, \mathcal{A}, L)$ where the parameter space of $\mathcal{M}$ is finite, the Bayesian estimator with any prior distribution does not satisfy the strict Condorcet criterion.*

Proof idea: The theorem is proved by contradiction. For the sake of contradiction, suppose the Bayesian estimator $BE_{\mathcal{F}}$ of $\mathcal{F}$ satisfies strict Condorcet criterion. Recall from Definition 3.22 that for any preference profile D, w_D is the weight function for the weighted majority graph of D. Then, the following lemma can be proved.

Lemma 7.22   *Suppose $BE_{\mathcal{F}}$ satisfies strict Condorcet criterion. For any profile D and any pair of alternatives (a, b), if $w_D(a \to b) = 0$ then $BL_{\mathcal{F}}(a, D) = BL_{\mathcal{F}}(b, D)$.*

Now, consider any profile D where $w_D(a \to b) = w_D(b \to c) = 0$, $w_D(a \to c) = 2$, a and b are the only two weak Condorcet winners, and c loses to all other alternatives in head-to-head competitions. Such a profile exists due to McGarvey's theorem [McGarvey, 1953]. By Lemma 7.22, $BL_{\mathcal{F}}(a, D) = BL_{\mathcal{F}}(b, D) = BL_{\mathcal{F}}(c, D)$. However, because $BE_{\mathcal{F}}$ satisfies the strict Condorcet criterion, $c \notin BE_{\mathcal{F}}(D)$, which is a contradiction. $\square$

We note that Theorem 7.21 works for any loss function in the framework and any prior over the parameter space. A direct corollary of Theorem 7.21 is that any voting rule that satisfies strict Condorcet criterion cannot be the Bayesian estimator of any framework with finite parameter space.

Corollary 7.23   Copeland$_1$, maximin, Black's function, Dodgson's function, Young's function, Condorcet's function, and Fishburn's function[3] cannot be the Bayesian estimator of any framework with finite parameter space.

7.3.2   SATISFIABILITY OF OTHER AXIOMS

In this subsection, we will examine the satisfaction of other axioms defined in Section 7.1.2 for two Bayesian estimators: $BE_{Ma,\varphi}^{Top}$ and $BE_{Co,\varphi}^{Borda}$ defined in Example 7.7. We will also see a positive result, saying that all these axioms can be satisfied by the Bayesian estimator of another

---

[3]Definitions of these rules except Copeland and maximin can be found in Fishburn [1977], where it was proved that they satisfy the strict Condorcet criterion.

natural framework, and additionally, the Bayesian estimator is minimax and can be computed in polynomial time. Recall that in the framework the decision space $\mathcal{DS}$ is set to be $\mathcal{A}$ to apply the axioms.

Let us start with neutrality. The following two theorems establish the equivalence between neutral Bayesian estimators and Bayesian estimators of neutral frameworks. For neutral frameworks, we further require that $\sigma_A = \sigma$ in Definition 7.15, where σ_A is the corresponding permutation over $\mathcal{A}$.

Theorem 7.24 [Xia, 2016].   *The Bayesian estimator of any neutral framework satisfies neutrality.*

Theorem 7.24 offers a sufficient condition for a Bayesian estimator to satisfy neutrality.

Theorem 7.25 [Xia, 2016].   *If the Bayesian estimator $BE_{\mathcal{F}}$ of a framework $\mathcal{F}$ satisfies neutrality, then there exists a neutral framework $\mathcal{F}^*$ such that $BE_{\mathcal{F}^*} = BE_{\mathcal{F}}$.*

Theorem 7.25 says that while it might be possible for $BE_{\mathcal{F}}$ of a non-neutral framework $\mathcal{F}$ to satisfy neutrality, there exists a neutral framework whose Bayesian estimator is equivalent to $BE_{\mathcal{F}}$. This means that when designing a neutral Bayesian estimator, it suffices to focus on Bayesian estimators of neutral frameworks. This can sometimes simplify the design problem, as the *revelation principle* does for mechanism design [Nisan et al., 2007].

Now let us define a new statistical model for rank data.

Definition 7.26   For any $0 < \varphi < 1$, let $\mathcal{M}_{\mathrm{Pair},\varphi}$ denote the following statistical model for rank data: The parameter space $\Theta = \{\theta_{bc} : b \neq c \in \mathcal{A}\}$. For any $R \in \mathcal{L}(\mathcal{A})$, we let $\pi_{\theta_{bc}}(R) \propto$ $\begin{cases} 1 & \text{if } b \succ_R c \\ \varphi & \text{otherwise} \end{cases}$. Let $L_1(\theta_{bc}, a) = \begin{cases} 1 & \text{if } a = c \\ 0 & \text{otherwise} \end{cases}$ and let $\mathcal{F}^1_{\mathrm{Pair},\varphi} = (\mathcal{M}_{\mathrm{Pair},\varphi}, \mathcal{A}, L_1)$. Let $BE^1_{\mathrm{Pair},\varphi}$ denote the Bayesian estimator of $\mathcal{F}^1_{\mathrm{Pair},\varphi}$ with the uniform prior.

That is, parameters in $\mathcal{M}_{\mathrm{Pair},\varphi}$ correspond to pairwise comparisons between alternatives. A parameter θ_{bc} can be interpreted as "$b \succ c$ is the strongest pairwise comparison." L_1 says that the loss of a is 1 w.r.t. θ_{bc} if and only if a is the less preferred alternative c.

The satisfaction of axioms in Section 7.1.2, minimaxity (Definition 7.9), and computational complexity for $BE^{\mathrm{Top}}_{\mathrm{Ma},\varphi}$, $BE^{\mathrm{Borda}}_{\mathrm{Co},\varphi}$, and $BE^1_{\mathrm{Pair},\varphi}$ are summarized in Table 7.2 [Azari Soufiani et al., 2014b, Procaccia et al., 2012, Xia, 2016].

We note that $BE^1_{\mathrm{Pair},\varphi}$ satisfies all axioms defined in Section 7.1.2 except the strict Condorcet Criterion, which cannot be satisfied by any Bayesian estimator of any framework with finite sample space (Theorem 7.21). Moreover, $BE^1_{\mathrm{Pair},\varphi}$ also satisfies minimaxity w.r.t. $\mathcal{F}^1_{\mathrm{Pair},\varphi}$ and can be computed in polynomial time.

Table 7.2: Comparisons of three Bayesian estimators. m is the number of alternatives. φ is the dispersion parameter [Azari Soufiani et al., 2014b, Procaccia et al., 2012, Xia, 2016].

	Anon.	Strict Cond.	Neu.	Mono.	Cond.	Mnmx	Comp.
$\text{BE}^{\text{Top}}_{\text{Ma},\varphi}$					Y iff $\dfrac{\varphi(1-\varphi^{m-1})}{1-\varphi} \leq 1$		NP-hard
$\text{BE}^{\text{Borda}}_{\text{Co},\varphi}$	Y	N (Thm. 7.20)	Y (Thm. 7.22)	Y	Y iff $\varphi \leq \dfrac{1}{m-1}$	Y	P
$\text{BE}^{1}_{\text{Pair},\varphi}$					Y iff $\varphi \leq \dfrac{1}{m-1}$		P

7.4 AN AUTOMATED DESIGN FRAMEWORK

Suppose we are designing a mechanism to rank products as in Scenario 6 in Chapter 1. Ideally, we would like the mechanism to be optimal or nearly optimal in the statistical sense as well as being socially desirable, for example being fair. Suppose a statistical decision-theoretic framework $\mathcal{F}$ has been built to describe users' preferences and the loss function. In addition, a set $\mathcal{X}$ of axioms are given. We are facing the following design problem.

The Design Problem. *How can we design a computationally tractable mechanism r that (approximately) minimizes Bayesian (or frequentist) loss in $\mathcal{F}$ while satisfying axioms in $\mathcal{X}$?*

This problem is already challenging for satisfying axioms in $\mathcal{X}$ alone, which is a central topic in modern social choice theory. Unfortunately, social choice theory does not offer a clear answer. Various impossibility theorems were proved, showing that some axioms are incompatible with others, yet mechanism design in social choice is still largely a manual practice—usually a mechanism is designed by hand with good intuitions, then its satisfaction of desirable axioms are analyzed and compared to that of other mechanisms. For example, see Schulze [2011] for a comparison of some existing mechanisms w.r.t. some axioms. Needless to say, the introduction of the statistical criteria and computational criteria to the design problem just makes it even more challenging.

This section presents some ideas to build an automated framework to solve the design problem, as illustrated in Figure 7.2.

In the automated design framework shown in Figure 7.2, the data D_{design} consist of $\{(D, d)\} \subseteq \mathcal{S} \times \mathcal{DS}$. That is, each data point is a (D, d) pair, where D is the rank data and

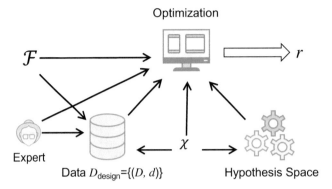

Figure 7.2: The automated design framework.

d is the decision that ought to be made for D. An optimization algorithm is performed on D_{design} to learn a decision mechanism r from the hypothesis space.

For the purpose of illustration, we will focus on decision mechanisms that output a single decision in this section. We will discuss how to incorporating $\mathcal{F}$, $\mathcal{X}$, and experts' inputs to D_{design} in Section 7.4.1; how to incorporate $\mathcal{X}$ to the hypothesis space in Section 7.4.2; and how to incorporate experts' inputs, $\mathcal{F}$, and $\mathcal{X}$ to the objective function for optimization in Section 7.4.3.

7.4.1   DATA GENERATION

In the automated design framework, the data D_{design} come from three sources.

- **Experts:** the mechanism designer can query human experts to suggest the desired decision d for any rank data D.

- **$\mathcal{F}$:** For any rank data D, an optimal decision d in the Bayesian sense or frequentist sense can be computed.

- **$\mathcal{X}$:** As we will see soon, many axioms in social choice can be viewed as data generation rules. Therefore, any axiom in $\mathcal{X}$ can be used to generate new datapoints from D_{design}, and then add them to D_{design}.

Social Choice Axioms as Data Generation Rules.   Most social choice axioms can be categorized into the following three classes. Each class represents a set of inference rules modeled as *Horn clauses*, where each (D, d) is a binary variable that takes 1 if $d = r(D)$; otherwise it takes 0.

- A *pointwise axiom* is characterized by a set of logical formulas $\{\bot \to (D, d)\}$, each of which means "the decision for D must be d." For example, the Condorcet criterion (Definition 7.12) is a pointwise axiom. A pointwise axiom can be seen as a human expert.

- A *pairwise axiom* is characterized by a set of logical formulas $\{(D_1, d_1) \to (D_2, d_2)\}$, each of which means "if the decision for D_1 is d_1, then the decision for D_2 is d_2." For example, anonymity (Definition 7.10), neutrality (Definition 7.11), and monotonicity (Definition 7.14) are pairwise axioms.

- A *triple-wise axiom* is characterized by a set of logical formulas $\{(D_1, d_1) \land (D_2, d_2) \to (D_3, d_3)\}$, each of which means "if the decision for D_1 is d_1 and the decision for D_2 is d_2, then the decision for D_3 is d_3." For example, *consistency* is a triple-wise axiom, which states that if $r(D_1) = r(D_2) = d$, then $r(D_1 \cup D_2) = d$.

7.4.2 HYPOTHESIS SPACE

Suppose the mechanism designer can choose a hypothesis space of decision mechanisms, each of which satisfies some axioms in $\mathcal{X}$, then it is guaranteed that the outcome of the automated framework in Figure 7.2 will also satisfy the same set of axioms.

In order for this natural idea to work, we need to have good mathematical understandings on the structures of decision mechanisms that satisfy a combination of axioms. Such results are rare, and are often know as *axiomatic characterizations* in social choice. For example, the class of *positional scoring rules*[4] are characterized by anonymity, neutrality, consistency, and *continuity* [Young, 1975]. Sample complexity of learning positional scoring rules was considered by Procaccia et al. [2009].

One reasonable choice of hypothesis space is *generalized decision scoring rules* [Xia, 2015, Xia and Conitzer, 2008], illustrated in Figure 7.3.

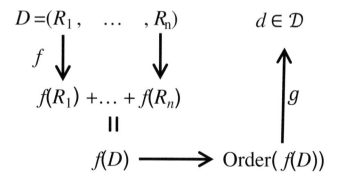

Figure 7.3: Generalized decision scoring rule [Xia, 2015, Xia and Conitzer, 2008].

A generalized decision scoring rule is defined by two functions: (1) $f : \mathcal{L}(\mathcal{A}) \to \mathbb{R}^l$, where $l \in \mathbb{N}$ is a given natural number and (2) g, which maps all weak orders over the l components of $\mathbb{R}^l$ to a decision $d \in \mathcal{DS}$. Each $f(R)$ is called a *generalized scoring vector*. Given any prefer-

---

[4]See Definition 6.7 for the randomized variant of positional scoring rules.

ence profile D, let $f(D) = \sum_{R \in D} f(R)$. Then, let $\text{Order}(f(D))$ denote the ordering over the l components of $f(D)$. Finally, the decision is $g(\text{Order}(f(D)))$.

It has been shown that many existing decision mechanisms are generalized decision scoring rules, and the class of generalized decision scoring rules are characterized by anonymity, *homogeneity*, and *finite local consistency* [Xia, 2015, Xia and Conitzer, 2009]. Moreover, generalized decision scoring rules are equivalent to combinations of linear binary classifiers via decision trees, for which many machine learning methods can be applied.

Example 7.27 STV as a GDSR. The *single transferable vote (STV)* voting rule is a GDSR, where $\mathcal{DSS} = \mathcal{A}$. STV selects the winner in the following $m - 1$ steps: in each step, the alternative ranked at the top positions least often is eliminated from the profile;[5] and the remaining alternative is the winner.

For STV, the generalized scoring vectors have exponentially many components. For every proper subset C of $\mathcal{A}$ and every alternative c not in C, there is a component in the vector that contains the number of times that c is ranked at the top if all alternatives in C are removed. The GDSR for STV is defined as follows.

- $l = \sum_{i=0}^{m-1} \binom{m}{i}(m - i)$; the elements of generalized scoring vectors are indexed by (C, c), where C is a proper subset of $\mathcal{A}$ and $c \notin C$.

- For any $R \in \mathcal{L}(\mathcal{A})$, $C \subsetneq \mathcal{A}$, and any $c \notin C$, the (C, c)-component of $f(R)$ is 1 if after removing C from R, c is ranked at the top; otherwise it is 0.

- g mimics the process of STV to select the winner.

7.4.3 OPTIMIZATION

Given D_{design} and a decision mechanism r, the following high-level objective function can be used for optimization:

$$\text{Obj}(r) = \alpha_E \cdot \text{Loss}_E\left(r, D_{\text{design}}\right) + \alpha_{\mathcal{F}} \cdot \text{Loss}_{\mathcal{F}}\left(r, D_{\text{design}}\right) + \alpha_{\mathcal{X}} \cdot \text{Loss}_{\mathcal{X}}\left(r, D_{\text{design}}\right),$$

where $\alpha_E, \alpha_{\mathcal{F}}$, and $\alpha_{\mathcal{X}}$ are non-negative weights, and

- $\text{Loss}_E(r, D_{\text{design}})$ represents the error r makes on human expert data;

- $\text{Loss}_{\mathcal{F}}(r, D_{\text{design}})$ represents the total loss of r in $\mathcal{F}$. For example, the Bayesian total loss of r can be defined as $\sum_{(D,c) \in D_{\text{design}}} \text{BL}_{\mathcal{F}}(r(D), D)$; and

- $\text{Loss}_{\mathcal{X}}(r, D_{\text{design}})$ represents the total loss of r w.r.t. it satisfaction of axioms in $\mathcal{X}$. For each $Ax \in \mathcal{X}$ represented by a logical formula defined in Section 7.4.1, its degree of satisfaction

---

[5]In case there is a tie, a tie-breaking mechanism applied.

under r can be evaluated by (randomly) checking $s + t$ instances, where s is the number of instances whose premises are true but the conclusion is false, and t is the number of formulas whose premises and the conclusion are all true. Then, the loss of r w.r.t. Ax can be approximated by $s/(s + t)$. $\mathrm{Loss}_{\mathcal{X}}(r, D_{\mathrm{design}})$ is the total loss of r w.r.t. all axioms in $\mathcal{X}$.

Overall, the automated design framework computes a decision mechanism r with minimum $\mathrm{Obj}(r)$.

7.5   BIBLIOGRAPHICAL NOTES

Section 7.1.   Berger [1985] offers a comprehensive survey on statistical decision theory. The statistical decision-theoretic framework for social choice was defined in Azari Soufiani et al. [2014b]. Commonly studied axioms and social choice mechanisms can be found in Nurmi [1987]. The axiomatic approach has been adopted in, for example, ranking systems [Altman and Tennenholtz, 2010], recommender systems [Pennock et al., 2000], and community detection [Borgs et al., 2016].

Section 7.2.   The section is based on Xia [2016]. Corollary 7.19 was proved by Caragiannis et al. [2016b] using different techniques under the sample complexity interpretation (see the discussion after Definition 7.9).

Section 7.3.   The section is largely based on Azari Soufiani et al. [2014b] and Xia [2016]. The NP-hardness of $\mathrm{BE}_{\mathrm{Ma},\varphi}^{\mathrm{Top}}$ in Table 7.2 was proved by Procaccia et al. [2012]. Conitzer and Sandholm [2005a] proved that the MLE of any statistical model must satisfy *consistency*. Pivato [2013] studied voting rules that can be viewed as Bayesian estimators. The interplay between statistical properties and other socio-economic properties, especially strategy-proofness, has been studied in Cai et al. [2015], Caragiannis et al. [2016a], Chen et al. [2018], Hardt et al. [2016], and Meir et al. [2010]. Some recent progresses are summarized in the book by Radanovic and Faltings [2017]. Other fairness in machine learning can be found in the survey by Kearns [2017].

Section 7.4.   The section is largely based on Xia [2013]. There is a large body of literature on automated mechanism design [Conitzer and Sandholm, 2002], which focuses on designing incentive compatible mechanisms.

CHAPTER 8

Future Directions

The three challenges and questions in Section 1.1 naturally lead to the following three directions.

Handling Richer Types of Rank Data   In previous work, it is commonly assumed that the data are composed of a particular type of ranks, for example choice sets, linear orders, or pairwise comparisons. A natural question arises when the assumption does not hold any more.

How can we model and learn from richer types of rank data?

While there are already EM-based algorithms for learning from general partial orders [Liu et al., 2019, Lu and Boutilier, 2014], the problem is more complicated than it appears. For example, suppose an agent says "I think both a and b are better than c". Then, there are at least following three possibilities for a vs. b (see Pini et al. [2007] for more discussions).

1. Indifference. That is, a and b are equally good. This case can be handled in RUMs by adding a soft constraint to require that the agent's latent utility of a is close to that of b.

2. Incompleteness. That is, it can be $a \succ b$ or $b \succ a$ but the agent did not specify which way it goes. This case can be handled by reasoning about uncertainty and analyzing whether the group decision is already determined regardless of how preferences are completed (called the *necessary winner* problem), or whether there is a way to complete the partial preferences so that a specific alternative becomes the group decision (called the *possible winner* problem) [Konczak and Lang, 2005].

3. Incomparability. That is, a and b are not comparable as the agent cannot compare apples to oranges. This case can be handled by augmenting the relationship between a and b with a third value, namely "incomparable." New statistical models need to be built and new group decision-making mechanisms need to be designed.

Similar challenges exist in many other situations. How can we deal with transitivity in partial orders? How can we model agents' confidence in their preferences? Should we ask agents to use ranks or numerical values to represent their preferences? Perhaps more importantly, but somewhat beyond the scope this book, how can we incorporate all sorts of ordinal and cardinal information to build a better model?

Smarter Interactions   As discussed in Chapter 6, the main research question for smarter and more effective interactions between the intelligent system and humans is the following.

How can the intelligent system obtain data more efficiently and effectively?

Both personal choice and social choice face the following challenges.

First, the standard approach following Bayesian experimental design focused on estimating the expected information gain myopically due to high computational cost. Consequently, how to go beyond this myopic paradigm is an important open question. See Section 6.5 for some recent progress.

Second, as we have seen in Section 6.2.1, various information criteria have been proposed as measures of quality of data. It would be useful to design a general framework that can automatically identify the best criterion, or the best combination of multiple criteria, or even sometimes proposing new criteria. Perhaps (deep) reinforcement learning can be leveraged to build the framework.

Third, how can the system be more human-aware by creating an immersive environment for decision-making? Currently, most research assumes that preferences are obtained by directly querying humans. A more natural and less intrusive mode of interaction would be that the system silently learns human preferences for most of the time from their behavior and conversations, and only queries humans when necessary. To achieve this goal, it would be useful to learn preferences and behavior from natural language, e.g., by leveraging existing NLP techniques. It is also important for the system to be aware of uncertainty in a single agent's preferences (for personal choice) as well as uncertainty in the consensus (for social choice). Definitions and techniques from *uncertainty quantification* [Smith, 2013] can be useful.

Multi-Criteria Design and Analysis   This direction is a natural extension of the main theme of the book, especially given richer rank data obtained by smarter interactions.

How can we design intelligent systems with desirable statistical, computational, and socio-economic properties?

Statistical aspects. One natural question is how to design and select statistical models that better fit data. We have seen comparisons of some popular models for Preflib data w.r.t. three popular model selection criteria in Section 2.1. Fortunately, all three criteria agree on the comparisons, but in general this may not happen. Perhaps model selection can be done by incorporating agents' feedbacks about the decisions, but such data are generally unavailable. Also, there is still a long way to go in applying principled statistical inference methods, such as the Bayesian principle and the frequentists' principle, to decision-making scenarios. In particular, not much work has been done under frequentists' principle, which can be viewed as robust decision-making.

Computational aspects. Designing faster algorithms to compute the decisions is always an important direction for research. For each algorithm, it is important to understand its sample complexity as well as its computational complexity.

Socio-Economic aspects. Following Section 7.3 and Table 7.2, one important open question is to build a comprehensive picture of tradeoffs among statistical, computational, and socio-economic properties, possibly in the form of impossibility theorems like Theorem 7.21. Practically, obtaining better mechanisms from the unified consideration of statistical, computational, and socio-economic criteria is an important and challenging next step.

Bibliography

Ehsan Abbasnejad, Scott Sanner, Edwin V. Bonilla, and Pascal Poupart. Learning community-based preferences via Dirichlet process mixtures of Gaussian processes. In *Proc. of the 23rd International Joint Conference on Artificial Intelligence*, pages 1213–1219, 2013. 30

Arpit Agarwal, Prathamesh Patil, and Shivani Agarwal. Accelerated spectral ranking. In *Proc. of the 35th International Conference on Machine Learning*, 2018. 50

Nir Ailon, Moses Charikar, and Alantha Newman. Aggregating inconsistent information: Ranking and clustering. *Journal of the ACM*, 55(5), no. 23, 2008. DOI: 10.1145/1411509.1411513 50

Hirotugu Akaike. A new look at the statistical model identification. *IEEE Transactions on Automatic Control*, 19(6):716–723, 1974. DOI: 10.1007/978-1-4612-1694-0_16 31

Alnur Ali and Marina Meila. Experiments with Kemeny ranking: What works when? *Mathematical Social Sciences*, 64(1):28–40, 2012. DOI: 10.1016/j.mathsocsci.2011.08.008 51

Elizabeth S. Allman, Catherine Matias, and John A. Rhodes. Identifiability of parameters in latent structure models with many observed variables. *The Annals of Statistics*, 37(6A):3099–3132, 2009. DOI: 10.1214/09-aos689 83, 84, 90

Alon Altman and Moshe Tennenholtz. An axiomatic approach to personalized ranking systems. *Journal of the ACM*, 57(4), no. 26, 2010. DOI: 10.1145/1734213.1734220 125

Kenneth Arrow. *Social Choice and Individual Values*, 2nd ed (1st ed. 1951). New Haven, Cowles Foundation, 1963. 7

Pranjal Awasthi, Avrim Blum, Or Sheffet, and Aravindan Vijayaraghavan. Learning mixtures of ranking models. In *Proc. of Advances in Neural Information Processing Systems*, pages 2609–2617, 2014. 27, 89, 90, 91

Hossein Azari Soufiani, David C. Parkes, and Lirong Xia. Random utility theory for social choice. In *Proc. of Advances in Neural Information Processing Systems*, pages 126–134, 2012. 44, 45, 46, 50

Hossein Azari Soufiani, William Chen, David C. Parkes, and Lirong Xia. Generalized method-of-moments for rank aggregation. In *Proc. of Advances in Neural Information Processing Systems*, 2013a. 35, 39, 50, 53, 56, 57, 58, 60, 61, 75

Hossein Azari Soufiani, David C. Parkes, and Lirong Xia. Preference elicitation for general random utility models. In *Proc. of Uncertainty in Artificial Intelligence*, 2013b. 99, 110

Hossein Azari Soufiani, David C. Parkes, and Lirong Xia. Computing parametric ranking models via rank-breaking. In *Proc. of the 31st International Conference on Machine Learning*, 2014a. 45, 47, 50, 58, 63, 75

Hossein Azari Soufiani, David C. Parkes, and Lirong Xia. Statistical decision theory approaches to social choice. In *Proc. of Advances in Neural Information Processing Systems*, 2014b. 32, 113, 120, 121, 125

Mark Bagnoli and Ted Bergstrom. Log-concave probability and its applications. *Economic Theory*, 26(2):445–469, 2005. DOI: 10.1007/3-540-29578-x_11 72

John Bartholdi, III, Craig Tovey, and Michael Trick. Voting schemes for which it can be difficult to tell who won the election. *Social Choice and Welfare*, 6:157–165, 1989. DOI: 10.1007/bf00303169 49, 50

James O. Berger. *Statistical Decision Theory and Bayesian Analysis*, 2nd ed. Springer, 1985. DOI: 10.1007/978-1-4757-4286-2 113, 118, 125

Steven Berry, James Levinsohn, and Ariel Pakes. Automobile prices in market equilibrium. *Econometrica*, 63(4):841–890, 1995. DOI: 10.2307/2171802 4

Nadja Betzler, Michael R. Fellows, Jiong Guo, Rolf Niedermeier, and Frances A. Rosamond. Fixed-parameter algorithms for Kemeny rankings. *Theoretical Computer Science*, 410:4554–4570, 2009. DOI: 10.1016/j.tcs.2009.08.033 51

Debarun Bhattacharjya and Jeffrey Kephart. Bayesian interactive decision support for multi-attribute problems with even swaps. In *Proc. of the 30th Conference Annual Conference on Uncertainty in Artificial Intelligence*, 2014. 4

Nancy E. Bockstael. The use of random utility in modeling rural health care demand: Discussion. *American Journal of Agricultural Economics*, 81(3):692–695, 1999. DOI: 10.2307/1244036 4

Christian Borgs, Jennifer Chayes, Adrian Marple, and Shang-Hua Teng. An axiomatic approach to community detection. In *Proc. of the ACM Conference on Innovations in Theoretical Computer Science*, 2016. DOI: 10.1145/2840728.2840748 125

Craig Boutilier. Computational decision support: Regret-based models for optimization and preference elicitation. In P. H. Crowley and T. R. Zentall, Eds., *Comparative Decision Making: Analysis and Support Across Disciplines and Applications*. Oxford University Press, 2013. DOI: 10.1093/acprof:oso/9780199856800.003.0041 110

George E. P. Box. Robustness in the strategy of scientific model building. In R. L. Launer and G. N. Wilkinson, Eds., *Robustness in Statistics*, pages 201–236, Academic Press, 1979. DOI: 10.1016/b978-0-12-438150-6.50018-2 9

Ralph Allan Bradley and Milton E. Terry. Rank analysis of incomplete block designs: I. The method of paired comparisons. *Biometrika*, 39(3/4):324–345, 1952. DOI: 10.2307/2334029 31

Felix Brandt, Vincent Conitzer, Ulle Endriss, Jerome Lang, and Ariel D. Procaccia, Eds. *Handbook of Computational Social Choice.* Cambridge University Press, 2016. DOI: 10.1017/cbo9781107446984 7

Róbert Busa-Fekete, Eyke Hüllermeier, and Balázs Szörényi. Preference-based rank elicitation using statistical models: The case of mallows. In *Proc. of the 31st International Conference on International Conference on Machine Learning*, pages 1071–1079, 2014. 110

Yang Cai, Constantinos Daskalakis, and Christos Papadimitriou. Optimum statistical estimation with strategic data sources. In *Proc. of the 28th Conference on Computational Learning Theory*, pages 280–296, 2015. 125

Ioannis Caragiannis, Ariel D. Procaccia, and Nisarg Shah. Truthful univariate estimators. In *Proc. of the 33rd International Conference on Machine Learning*, pages 127–135, 2016a. 125

Ioannis Caragiannis, Ariel D. Procaccia, and Nisarg Shah. When do noisy votes reveal the truth? *ACM Transactions on Economics and Computation*, 4(3), no. 15, 2016b. DOI: 10.1145/2492002.2482570 125

George Casella and Roger L. Berger. *Statistical Inference*, 2nd ed. Cengage Learning, 2001. DOI: 10.2307/2290879 33

Urszula Chajewska, Daphne Koller, and Ron Parr. Making rational decisions using adaptive utility elicitation. In *Proc. of the National Conference on Artificial Intelligence*, pages 363–369, 2000. 109

Kathryn Chaloner and Isabella Verdinelli. Bayesian experimental design: A review. *Statistical Science*, 10(3):273–304, 1995. DOI: 10.1214/ss/1177009939 94, 96, 109

Yiling Chen, Chara Podimata, Ariel D. Procaccia, and Nisarg Shah. Strategyproof linear regression in high dimensions. In *Proc. of the 19th ACM Conference on Economics and Computation*, 2018. DOI: 10.1145/3219166.3219175 125

Flavio Chierichetti, Anirban Dasgupta, Ravi Kumar, and Silvio Lattanzi. On learning mixture models for permutations. In *Proc. of the ACM Conference on Innovations in Theoretical Computer Science*, 2015. DOI: 10.1145/2688073.2688111 91

Flavio Chierichetti, Ravi Kumar, and Andrew Tomkins. Learning a mixture of two multinomial logits. In *Proc. of the 35th International Conference on Machine Learning*, 2018. 90

Marquis de Condorcet. *Essai sur L'application de L'analyse à la Probabilité des Décisions Rendues à la Pluralité des Voix*. L'Imprimerie Royale, Paris, 1785. DOI: 10.1017/cbo9781139923972.002 30, 112

Vincent Conitzer and Tuomas Sandholm. Complexity of mechanism design. In *Proc. of the 18th Annual Conference on Uncertainty in Artificial Intelligence*, pages 103–110, 2002. 125

Vincent Conitzer and Tuomas Sandholm. Common voting rules as maximum likelihood estimators. In *Proc. of the 21st Annual Conference on Uncertainty in Artificial Intelligence*, pages 145–152, 2005a. 125

Vincent Conitzer and Tuomas Sandholm. Communication complexity of common voting rules. In *Proc. of the ACM Conference on Electronic Commerce*, pages 78–87, 2005b. DOI: 10.1145/1064009.1064018 110

Vincent Conitzer, Andrew Davenport, and Jayant Kalagnanam. Improved bounds for computing Kemeny rankings. In *Proc. of the National Conference on Artificial Intelligence*, pages 620–626, 2006. 49, 50

Harald Cramér. *Mathematical Methods of Statistics*. Princeton University Press, 1946. DOI: 10.1515/9781400883868 74

Andreas Darmann, Edith Elkind, Sascha Kurz, Jérôme Lang, Joachim Schauer, and Gerhard Woeginger. Group activity selection problem. In *Proc. of the 8th International Conference on Internet and Network Economics*, pages 156–169, 2012. DOI: 10.1007/978-3-642-35311-6_12 2

A. P. Dempster, N. M. Laird, and D. B. Rubin. Maximum likelihood from incomplete data via the EM algorithm. *Journal of the Royal Statistical Society. Series B*, 39(1):1–38, 1977. DOI: 10.1111/j.2517-6161.1977.tb01600.x 41

Antoine Desir, Vineet Goyal, Srikanth Jagabathula, and Danny Segev. Assortment optimization under the Mallows model. In *Proc. of the 30th International Conference on Neural Information Processing Systems*, pages 4707–4715, 2016. 91

Persi Diaconis and R. L. Graham. Spearman's footrule as a measure of disarray. *Journal of the Royal Statistical Society. Series B*, 39(2):262–268, 1977. DOI: 10.1111/j.2517-6161.1977.tb01624.x 32

Persi Diaconis and Phil Hanlon. Eigenanalysis for some examples of the metropolis algorithm. *Contemporary Mathematics*, 138:99–117, 1992. DOI: 10.1090/conm/138/1199122 32

Jean-Paul Doignon, Aleksandar Pekeč, and Michel Regenwetter. The repeated insertion model for rankings: Missing link between two subset choice models. *Psychometrika*, 69(1):33–54, 2004. DOI: 10.1007/bf02295838 28, 32

Joanna Drummond and Craig Boutilier. Preference elicitation and interview minimization in stable matchings. In *Proc. of the National Conference on Artificial Intelligence*, pages 645–653, 2014. 109

Cynthia Dwork, Ravi Kumar, Moni Naor, and D. Sivakumar. Rank aggregation methods for the Web. In *Proc. of the 10th World Wide Web Conference*, pages 613–622, 2001. DOI: 10.1145/371920.372165 4

Sylvain Ehrenfeld. On the efficiency of experimental designs. *The Annals of Mathematical Statistics*, 26(2):247–255, 1955. DOI: 10.1214/aoms/1177728541 98

Edith Elkind and Nisarg Shah. Electing the most probable without eliminating the irrational: Voting over intransitive domains. In *Proc. of the 30th Conference on Uncertainty in Artificial Intelligence*, pages 182–191, 2014. 32

Seda Erdem and Danny Campbell. Preferences for public involvement in health service decisions: A comparison between best-worst scaling and trio-wise stated preference elicitation techniques. *The European Journal of Health Economics*, 18(9):1107–1123, 2017. DOI: 10.1007/s10198-016-0856-4 109

Peter C. Fishburn. Condorcet social choice functions. *SIAM Journal on Applied Mathematics*, 33(3):469–489, 1977. DOI: 10.1137/0133030 112, 119

Michael Fligner and Joseph Verducci. Distance based ranking models. *Journal of the Royal Statistical Society B*, 48:359–369, 1986. DOI: 10.1111/j.2517-6161.1986.tb01420.x 32

Lester R. Ford, Jr. Solution of a ranking problem from binary comparisons. *The American Mathematical Monthly*, 64(8):28–33, 1957. DOI: 10.2307/2308513 50

Andrew Goett, Kathleen Hudson, and Kenneth E. Train. Customer choice among retail energy suppliers. *Energy Journal*, 21(4):1–28, 2002. 4

Isobel Claire Gormley and Thomas Brendan Murphy. Exploring voting blocs within the Irish exploring voting blocs within the Irish electorate: A mixture modeling approach. *Journal of the American Statistical Association*, 103(483):1014–1027, 2008. 79, 86

Alastair R. Hall. *Generalized Method of Moments*. Oxford University Press, 2005. DOI: 10.1002/9780470061602.eqf19004 87, 88

Lars Peter Hansen. Large sample properties of generalized method of moments estimators. *Econometrica*, 50(4):1029–1054, 1982. DOI: 10.2307/1912775 38, 40, 50, 57

Moritz Hardt, Nimrod Megiddo, Christos Papadimitriou, and Mary Wootters. Strategic classification. In *Proc. of the 7th Innovations in Theoretical Computer Science Conference*, pages 111–122, 2016. DOI: 10.1145/2840728.2840730 125

F. Maxwell Harper and Joseph A. Konstan. The MovieLens datasets: History and context. *ACM Transactions on Interactive Intelligent Systems*, 5(4), no. 19, 2015. DOI: 10.1145/2827872 30

Edith Hemaspaandra, Holger Spakowski, and Jörg Vogel. The complexity of Kemeny elections. *Theoretical Computer Science*, 349(3):382–391, December 2005. DOI: 10.1016/j.tcs.2005.08.031 50

John J. Horton. The dot-guessing game: A "fruit fly" for human computation research. *Human Computation Research*, 2010. DOI: 10.2139/ssrn.1600372 1

Dongling Huang and Lan Luo. Consumer preference elicitation of complex products using fuzzy support vector machine active learning. *Marketing Science*, 35(3):445–464, 2016. DOI: 10.1287/mksc.2015.0946 109

Eyke Hüllermeier, Johannes Fürnkranz, Weiwei Cheng, and Klaus Brinker. Label ranking by learning pairwise preferences. *Artificial Intelligence*, 172:1897–1916, 2008. DOI: 10.1016/j.artint.2008.08.002 30

David R. Hunter. MM algorithms for generalized Bradley-Terry models. *The Annals of Statistics*, vol. 32, pages 384–406, 2004. DOI: 10.1214/aos/1079120141 34, 35, 36, 37, 50

Clifford M. Hurvich and Chih-Ling Tsai. Regression and time series model selection in small samples. *Biometrika*, 76(2):297–307, 1989. DOI: 10.2307/2336663 31

Shali Jiang, Gustavo Malkomes, Geoff Converse, Alyssa Shofner, Benjamin Moseley, and Roman Garnett. Efficient nonmyopic active search. In *Proc. of the 34th International Conference on Machine Learning*, pages 1714–1723, 2017. 109

Richard Karp. Reducibility among combinatorial problems. In Raymond E. Miller and James W. Thatcher, Eds., *Complexity of Computer Computations*, pages 85–103, Plenum Press, NY, 1972. DOI: 10.1007/978-1-4684-2001-2_9 48

Michael Kearns. Fair algorithms for machine learning. In *Proc. of the ACM Conference on Economics and Computation*, pages 1–1, 2017. DOI: 10.1145/3033274.3084096 125

Claire Kenyon-Mathieu and Warren Schudy. How to rank with few errors: A PTAS for weighted feedback arc set on tournaments. In *Proc. of the 39th Annual ACM Symposium on Theory of Computing*, pages 95–103, San Diego, CA, 2007. 51

Ashish Khetan and Sewoong Oh. Data-driven rank breaking for efficient rank aggregation. *Journal of Machine Learning Research*, 17(193):1–54, 2016. 64, 65, 70, 75

Tamara G. Kolda and Brett W. Bader. Tensor decompositions and applications. *SIAM Review*, 51(3):455–500, 2009. DOI: 10.1137/07070111x 91

Kathrin Konczak and Jérôme Lang. Voting procedures with incomplete preferences. In *Multidisciplinary Workshop on Advances in Preference Handling*, 2005. 127

Joseph B. Kruskal. Three-way arrays: Rank and uniqueness of trilinear decompositions, with application to arithmetic complexity and statistics. *Linear Algebra and its Applications*, 18(2):95–138, 1977. DOI: 10.1016/0024-3795(77)90069-6 85

David A. Levin, Yuval Peres, and Elizabeth L. Wilmer. *Markov Chains and Mixing Times*. American Mathematical Society, 2008. DOI: 10.1090/mbk/058 58

Bruce G. Lindsay. Composite likelihood methods. *Contemporary Mathematics*, 80:220–239, 1988. DOI: 10.1090/conm/080/999014 102

Allen Liu and Ankur Moitra. Efficiently learning mixtures of Mallows models. In *Proc. of the 59th Annual IEEE Symposium on Foundations of Computer Science*, 2018. DOI: 10.1109/focs.2018.00066 91

Ao Liu, Zhibing Zhao, Chao Liao, Pinyan Lu, and Lirong Xia. Learning Plackett-Luce mixtures from partial preferences. In *Proc. of AAAI*, 2019. 127

Tie-Yan Liu. *Learning to Rank for Information Retrieval*. Springer, 2011. DOI: 10.1007/978-3-642-14267-3 3, 7

Benedikt Loepp, Tim Hussein, and Jüergen Ziegler. Choice-based preference elicitation for collaborative filtering recommender systems. In *Proc. of the SIGCHI Conference on Human Factors in Computing Systems*, pages 3085–3094, ACM, 2014. DOI: 10.1145/2556288.2557069 109

Tyler Lu and Craig Boutilier. Effective sampling and learning for Mallows models with pairwise-preference data. *Journal of Machine Learning Research*, 15:3963–4009, 2014. 32, 91, 127

R. Duncan Luce. The choice axiom after twenty years. *Journal of Mathematical Psychology*, 15(3):215–233, 1977. DOI: 10.1016/0022-2496(77)90032-3 31

Colin L. Mallows. Non-null ranking model. *Biometrika*, 44(1/2):114–130, 1957. DOI: 10.2307/2333244 26

John I. Marden. *Analyzing and Modeling Rank Data*. Chapman & Hall, 1995. DOI: 10.1201/b16552 32

George Marsaglia and Wai Wan Tsang. The Ziggurat method for generating random variables. *Journal of Statistical Software*, 5(8), 2000. DOI: 10.18637/jss.v005.i08 21, 23

Nicholas Mattei and Toby Walsh. PrefLib: A library of preference data. In *Proc. of 3rd International Conference on Algorithmic Decision Theory*, Lecture Notes in Artificial Intelligence, 2013. 30

Lucas Maystre and Matthias Grossglauser. Fast and accurate inference of Plackett-Luce models. In *Proc. of the 28th International Conference on Neural Information Processing Systems*, pages 172–180, 2015. 35, 36, 38, 50

David C. McGarvey. A theorem on the construction of voting paradoxes. *Econometrica*, 21(4):608–610, 1953. DOI: 10.2307/1907926 119

Geoffrey McLachlan and David Peel. *Finite Mixture Models*. John Wiley & Sons, 2004. DOI: 10.1002/0471721182 81, 90

Geoffrey J. McLachlan and Kaye E. Basford. *Mixture Models: Inference and Applications to Clustering*. Marcel Dekker, 1988. DOI: 10.2307/2348072 77

Reshef Meir, Ariel D. Procaccia, and Jeffrey S. Rosenschein. On the limits of dictatorial classification. In *Proc. of the 9th International Conference on Autonomous Agents and Multiagent Systems*, pages 609–616, 2010. 125

Michael Mitzenmacher and Eli Upfal. *Probability and Computing: Randomized Algorithms and Probabilistic Analysis*, 2nd ed. Cambridge University Press, 2017. DOI: 10.1017/cbo9780511813603 11

Cristina Mollica and Luca Tardella. Bayesian Plackett–Luce mixture models for partially ranked data. *Psychometrika*, pages 1–17, 2016. DOI: 10.1007/s11336-016-9530-0 91

Alexander M. Mood et al. On Hotelling's weighing problem. *The Annals of Mathematical Statistics*, 17(4):432–446, 1946. DOI: 10.1214/aoms/1177730883 98

Noam Nisan, Tim Roughgarden, Eva Tardos, and Vijay V. Vazirani. *Algorithmic Game Theory*. Cambridge University Press, New York, 2007. DOI: 10.1017/cbo9780511800481 120

Hannu Nurmi. *Comparing Voting Systems*. Springer, 1987. DOI: 10.1007/978-94-009-3985-1 125

Sewoong Oh and Devavrat Shah. Learning mixed multinomial logit model from ordinal data. In *Proc. of Advances in Neural Information Processing Systems*, pages 595–603, 2014. 90

Francesco Pauli, Walter Racugno, and Laura Ventura. Bayesian composite marginal likelihoods. *Statistica Sinica*, 21(1):149–164, 2011. 102, 103

Judea Pearl. *Probabilistic Reasoning in Intelligent Systems: Networks of Plausible Inference*. Morgan Kaufmann Publishers Inc., 1988. DOI: 10.1016/C2009-0-27609-4 18

R. N. Pendergrass and R. A. Bradley. Ranking in triple comparisons. In Ingram Olkin, Ed., *Contributions to Probability and Statistics*, pages 331–351, Stanford University Press, 1960. 32

David M. Pennock, Eric Horvitz, and C. Lee Giles. Social choice theory and recommender systems: Analysis of the axiomatic foundations of collaborative filtering. In *Proc. of the National Conference on Artificial Intelligence*, pages 729–734, 2000. 125

Thomas Pfeiffer, Xi Alice Gao, Andrew Mao, Yiling Chen, and David G. Rand. Adaptive polling and information aggregation. In *Proc. of the National Conference on Artificial Intelligence*, pages 122–128, 2012. 1, 2

Maria Silvia Pini, Francesca Rossi, Kristen Brent Venable, and Toby Walsh. Incompleteness and incomparability in preference aggregation. In *Proc. of the 20th International Joint Conference on Artificial Intelligence (IJCAI)*, Hyderabad, India, 2007. DOI: 10.1016/j.artint.2010.11.009 127

Marcus Pivato. Voting rules as statistical estimators. *Social Choice and Welfare*, 40(2):581–630, 2013. DOI: 10.1007/s00355-011-0619-1 125

Robin L. Plackett. The analysis of permutations. *Journal of the Royal Statistical Society. Series C (Applied Statistics)*, 24(2):193–202, 1975. DOI: 10.2307/2346567 31

Ariel D. Procaccia, Aviv Zohar, Yoni Peleg, and Jeffrey S. Rosenschein. The learnability of voting rules. *Artificial Intelligence*, 173:1133–1149, 2009. DOI: 10.1016/j.artint.2009.03.003 123

Ariel D. Procaccia, Sashank J. Reddi, and Nisarg Shah. A maximum likelihood approach for selecting sets of alternatives. In *Proc. of the 28th Conference on Uncertainty in Artificial Intelligence*, 2012. 120, 121, 125

Tao Qin and Tie-Yan Liu. Introducing LETOR 4.0 datasets. *ArXiv Preprint ArXiv:1306.2597*, 2013. 30

Goran Radanovic and Boi Faltings. *Game Theory for Data Science: Eliciting Truthful Information*. Morgan & Claypool Publishers, 2017. DOI: 10.2200/s00788ed1v01y201707aim035 125

Calyampudi Radakrishna Rao. Information and the accuracy attainable in the estimation of statistical parameters. *Bulletin of the Calcutta Mathematical Society*, 37:81–89, 1945. DOI: 10.1007/978-1-4612-0919-5_16 74

Baharak Rastegari, Paul Goldberg, and David Manlove. Preference elicitation in matching markets via interviews: A study of offline benchmarks. In *Proc. of the International Conference on Autonomous Agents and Multiagent Systems*, pages 1393–1394, 2016. 110

Tuomas Sandholm and Craig Boutilier. Preference elicitation in combinatorial auctions. In Peter Cramton, Yoav Shoham, and Richard Steinberg, Eds., *Combinatorial Auctions*, chapter 10, pages 233–263, MIT Press, 2006. DOI: 10.7551/mitpress/9780262033428.003.0011 110

Markus Schulze. A new monotonic, clone-independent, reversal symmetric, and condorcet-consistent single-winner election method. *Social Choice and Welfare*, 36(2):267–303, 2011. DOI: 10.1007/s00355-010-0475-4 121

Gideon Schwarz. Estimating the dimension of a model. *Annals of Statistics*, 6(2):461–464, 1978. DOI: 10.1214/aos/1176344136 31

Burr Settles. *Active Learning*. Morgan & Claypool Publishers, 2012. DOI: 10.2200/s00429ed1v01y201207aim018 109

Ralph C. Smith. *Uncertainty Quantification: Theory, Implementation, and Applications*. SIAM, 2013. 128

Peter Stone, Rodney Brooks, Erik Brynjolfsson, Ryan Calo, Oren Etzioni, Greg Hager, Julia Hirschberg, Shivaram Kalyanakrishnan, Ece Kamar, Sarit Kraus, Kevin Leyton-Brown, David Parkes, William Press, AnnaLee Saxenian, Julie Shah, Milind Tambe, and Astro Teller. Artificial intelligence and life in 2030. *One Hundred Year Study on Artificial Intelligence: Report of the 2015–2016 Study Panel*, Stanford University, Stanford, CA, September 2016. 1

David B. Thomas, Philip H. W. Leong, Wayne Luk, and John D. Villasenor. Gaussian random number generators. *ACM Computing Surveys*, 39(4):1–38, 2007. DOI: 10.1145/1287620.1287622 32

Louis Leon Thurstone. A law of comparative judgement. *Psychological Review*, 34(4):273–286, 1927. DOI: 10.1037/0033-295x.101.2.266 14, 31

Kenneth E. Train. *Discrete Choice Methods with Simulation*, 2nd ed. Cambridge University Press, 2009. 3, 7, 24, 32

Abraham Wald. On the efficient design of statistical investigations. *The Annals of Mathematical Statistics*, 14(2):134–140, 1943. DOI: 10.1214/aoms/1177731454 98

Shuaiqiang Wang, Shanshan Huang, Tie-Yan Liu, Jun Ma, Zhumin Chen, and Jari Veijalainen. Ranking-oriented collaborative filtering: A listwise approach. *ACM Transactions on Information Systems*, 35(2), no. 10, 2016. DOI: 10.1145/2960408 4

Marieke G. M. Weernink, Sarah I. M. Janus, Janine A. van Til, Dennis W. Raisch, Jeannette G. van Manen, and Maarten J. IJzerman. A systematic review to identify the use of preference elicitation methods in healthcare decision making. *Pharmaceutical Medicine*, 28(4):175–185, 2014. DOI: 10.1007/s40290-014-0059-1 109

Lirong Xia. Designing social choice mechanisms using machine learning. In *Proc. of the International Conference on Autonomous Agents and Multi-Agent Systemsulti-agent Systems*, pages 471–474, 2013. 125

Lirong Xia. Generalized decision scoring rules: Statistical, computational, and axiomatic properties. In *Proc. of the 16th ACM Conference on Economics and Computation*, pages 661–678, Portland, OR, 2015. DOI: 10.1145/2764468.2764518 123, 124

Lirong Xia. Bayesian estimators as voting rules. In *Proc. of the 32nd Conference on Uncertainty in Artificial Intelligence*, pages 785–794, 2016. 117, 119, 120, 121, 125

Lirong Xia and Vincent Conitzer. Generalized scoring rules and the frequency of coalitional manipulability. In *Proc. of the ACM Conference on Electronic Commerce*, pages 109–118, 2008. DOI: 10.1145/1386790.1386811 123

Lirong Xia and Vincent Conitzer. Finite local consistency characterizes generalized scoring rules. In *Proc. of the 21st International Joint Conference on Artificial Intelligence*, pages 336–341, 2009. 124

Rong Yang, Christopher Kiekintveld, Fernando Ordóñez, Milind Tambe, and Richard John. Improving resource allocation strategy against human adversaries in security games. In *Proc. of the 22nd International Joint Conference on Artificial Intelligence*, 2011. DOI: 10.1016/j.artint.2012.11.004 4

H. Peyton Young. Social choice scoring functions. *SIAM Journal on Applied Mathematics*, 28(4):824–838, 1975. DOI: 10.1137/0128067 123

H. Peyton Young. Condorcet's theory of voting. *American Political Science Review*, 82:1231–1244, 1988. DOI: 10.2307/1961757 30, 112, 115

Zhibing Zhao and Lirong Xia. Composite marginal likelihood methods for random utility models. In *Proc. of the 35th International Conference on Machine Learning*, 2018. 68, 70, 71, 72, 73, 75

Zhibing Zhao, Peter Piech, and Lirong Xia. Learning mixtures of Plackett-Luce models. In *Proc. of the 33rd International Conference on Machine Learning*, 2016. 32, 81, 82, 83, 87, 90

Zhibing Zhao, Haoming Li, Junming Wang, Jeffrey Kephart, Nicholas Mattei, Hui Su, and Lirong Xia. A cost-effective framework for preference elicitation and aggregation. In *Proc. of Uncertainty in Artificial Intelligence*, 2018a. 105, 106, 107, 108, 109, 110

Zhibing Zhao, Tristan Villamil, and Lirong Xia. Learning mixtures of random utility models. In *Proc. of the National Conference on Artificial Intelligence*, 2018b. 31, 32, 87, 88, 91

Author's Biography

LIRONG XIA

Dr. Xia is an associate professor in the Department of Computer Science at Rensselaer Polytechnic Institute (RPI). Prior to joining RPI in 2013, he was a CRCS fellow and NSF CI Fellow at the Center for Research on Computation and Society at Harvard University. He received his Ph.D. in Computer Science and M.A. in Economics from Duke University, and his B.E. in Computer Science and Technology from Tsinghua University. His research focuses on the intersection of computer science and microeconomics.

Dr. Xia is the recipient of an NSF CAREER award, a Simons-Berkeley Research Fellowship, the 2018 Rensselaer James M. Tien'66 Early Career Award, and was named as one of "AI's 10 to watch" by IEEE Intelligent Systems in 2015.

Printed in the United States
by Baker & Taylor Publisher Services